Ryanodine Receptors

Pharmacology and Toxicology: Basic and Clinical Aspects

Mannfred A. Hollinger, Series Editor
University of California, Davis

Published Titles

Inflammatory Cells and Mediators in Bronchial Asthma, 1990, Devendra K. Agrawal and Robert G. Townley
Pharmacology of the Skin, 1991, Hasan Mukhtar
In Vitro *Methods of Toxicology*, 1992, Ronald R. Watson
Basis of Toxicity Testing, 1992, Donald J. Ecobichon
Human Drug Metabolism from Molecular Biology to Man, 1992, Elizabeth Jeffreys
Platelet Activating Factor Receptor: Signal Mechanisms and Molecular Biology, 1992, Shivendra D. Shukla
Biopharmaceutics of Ocular Drug Delivery, 1992, Peter Edman
Beneficial and Toxic Effects of Aspirin, 1993, Susan E. Feinman
Preclinical and Clinical Modulation of Anticancer Drugs, 1993, Kenneth D. Tew, Peter Houghton, and Janet Houghton
Peroxisome Proliferators: Unique Inducers of Drug-Metabolizing Enzymes, 1994, David E. Moody
Angiotensin II Receptors, Volume I: Molecular Biology, Biochemistry, Pharmacology, and Clinical Perspectives, 1994, Robert R. Ruffolo, Jr.
Angiotensin II Receptors, Volume II: Medicinal Chemistry, 1994, Robert R. Ruffolo, Jr.
Chemical and Structural Approaches to Rational Drug Design, 1994, David B. Weiner and William V. Williams
Biological Approaches to Rational Drug Design, 1994, David B. Weiner and William V. Williams
Direct Allosteric Control of Glutamate Receptors, 1994, M. Palfreyman, I. Reynolds, and P. Skolnick
Genomic and Non-Genomic Effects of Aldosterone, 1994, Martin Wehling
Human Growth Hormone Pharmacology: Basic and Clinical Aspects, 1995, Kathleen T. Shiverick and Arlan Rosenbloom
Placental Toxicology, 1995, B. V. Rama Sastry
Stealth Liposomes, 1995, Danilo Lasic and Frank Marti
TAXOL®: Science and Applications, 1995, Matthew Suffness
Endothelin Receptors: From the Gene to the Human, 1995, Robert R. Ruffolo, Jr.
Serotonin and Gastrointestinal Function, 1995, Timothy S. Gaginella and James J. Galligan
Drug Delivery Systems, 1995, Vasant V. Ranade and Mannfred A. Hollinger
Experimental Models of Mucosal Inflammation, 1996, Timothy S. Gaginella
Ryanodine Receptors, 1996, Vincenzo Sorrentino

Pharmacology and Toxicology: Basic and Clinical Aspects

Mannfred A. Hollinger, Series Editor
University of California, Davis

Forthcoming Titles

Alcohol Consumption, Cancer and Birth Defects: Mechanisms Involved in Increased Risk Associated with Drinking, Anthony J. Garro
Alternative Methodologies for the Safety Evaluation of Chemicals in the Cosmetic Industry, Nicola Loprieno
Antibody Therapeutics, William J. Harris and John R. Adair
Brain Mechanisms and Psychotropic Drugs, A. Baskys and G. Remington
Chemoattractant Ligands and Their Receptors, Richard Horuk
CNS Injuries: Cellular Responses and Pharmacological Strategies, Martin Berry and Ann Logan
Immunopharmaceuticals, Edward S. Kimball
Muscarinic Receptor Subtypes in Smooth Muscle, Richard M. Eglen
Neural Control of Airways, Peter J. Barnes
Pharmacological Effects of Ethanol on the Nervous System, Richard A. Deitrich
Pharmacological Regulation of Gene Expression in the CNS, Kalpana Merchant
Pharmacology in Exercise and Sport, Satu M. Somani
Pharmacology of Intestinal Secretion, Timothy S. Gaginella
Phospholipase A_2 in Clinical Inflammation: Endogenous Regulation and Pathophysiological Actions, Keith B. Glaser and Peter Vadas
Placental Pharmacology, B. V. Rama Sastry
Receptor Characterization and Regulation, Devendra K. Agrawal
Receptor Dynamics in Neural Development, Christopher A. Shaw
Targeted Delivery of Imaging Agents, Vladimir P. Torchilin
Therapeutic Modulation of Cytokines, M.W. Bodmer and Brian Henderson

Ryanodine Receptors

Vincenzo Sorrentino

CRC Press
Boca Raton New York London Tokyo

Library of Congress Cataloging-in Publication Date

Ryanodine receptors / edited by Vincenzo Sorrentino.
p. cm. — (Pharmacology and toxicology)
Includes bibliographical references and index.
ISBN 0-8493-8543-1
1. Calcium channels. 2. Ryanodine—Receptors. I. Sorrentino, Vincenzo. II. Series: Pharmacology & toxicology (Boca Raton, Fla.)
QP535.C2R88 1995
574.87′5—dc20 95-14214
CIP

Direct all inquiries to CRC Press, Inc., 2000 Corporate Blvd., N.W., Boca Raton, Florida 33431.

International Standard Book Number 0-8493-8543-1
Library of Congress Card Number 95-14214
Printed in the United States of America 1 2 3 4 5 6 7 8 9 0
Printed on acid-free paper

SERIES PREFACE

The series, *Pharmacology and Toxicology: Basic and Clinical Aspects*, has been created in recognition of the fact that, from time to time, a new area of interest within a discipline matures to a critical mass that merits organization and integration of the respective observations into a free-standing monograph. In order for such an undertaking to be successful, each editor and/or author must be qualified to identify and select sources best suited to communicate essential aspects of that subject. Such is the case with *Ryanodine Receptors*. Vincenzo Sorrentino has assembled a list of international contributors whose expertise in their respective areas is well known.

Mannfred A. Hollinger, Ph.D.
Series Editor
Professor
Department of Medical Pharmacology and Toxicology
University of California, Davis
Davis, California

PREFACE

For scientists interested in understanding the mechanisms that regulate the release of calcium from intracellular stores, ryanodine receptors (RyRs)/calcium channels have long been a familiar name.

In the past, however, much attention was given to RyRs in the context of skeletal and cardiac muscle cells. The importance of these channels also for events occurring in other cell types such as neurons has begun to be recognized only recently following identification of RyRs in the central nervous system and other cell types. Therefore investigators from fields other than muscle biochemistry and physiology have turned their attention to RyRs. In fact, in addition to their fundamental role in the regulation of muscle contraction in skeletal and cardiac muscle, other physiological processes in which RyRs are involved seem to range from insulin release in pancreatic islets to long-term potentiation in the brain. A clinical relevance of these channels became manifest after the discovery of mutations in the RyR1 gene in individuals with malignant hyperthermia and central core disease.

Developments in the field of RyRs have accumulated at a relatively high pace in the last few years, so that this seems an opportune time to present a volume in which relevant information on this topic could be grouped. The first chapter, by Franzini-Armstrong, is a comprehensive review from an ultrastructural point of view of RyRs as components of muscle fibers, their location in the sarcoplasmic reticulum, and interaction with the dihydropyridine receptors. Chapters 2 through 5 deal with the pharmacological regulation of RyRs. Chapter 2 by Ehrlich and Marks starts by introducing the different RyR isoforms and then provides an overview on the regulation of the calcium channel activity. Chapter 3 by Hon Cheung Lee and Chapter 4 by Galione and Summerhill extensively cover cyclic ADP-ribose, beginning with its structure and metabolic pathways and including a discussion of present evidence that makes cyclic ADP-ribose, both in sea urchin and in mammalian cells, an important agonist of calcium-induced calcium release through RyRs. Morrissette and Coronado discuss in Chapter 5 the specific pharmacological properties of different RyR isoforms and present the results of studies based on RyR-specific scorpion toxins that have proven to be useful tools in dissecting the channel structure and function.

Chapter 6 reviews the work done in several labs on the molecular biology of RyRs from the initial cloning of RyR cDNAs and identification of potential regulatory domains from analysis of the deduced amino acid sequence to chromosomal localization of the genes and knockout of the RyR1 in mouse.

Chapter 7 by Nori, Gorza, and Volpe reviews all available information on the expression of the three known RyR isoforms, both in tissues and cells, and introduces the potential importance of coexpression of RyRs and inositol-1,4,5- trisphosphate receptors for intracellular calcium homeostasis. Berridge and colleagues, in Chapter 8, elegantly discuss the role of RyRs in signaling in nonmuscle cells — an area that has only very recently started to develop, but that may provide even more surprises in the future. The book ends with the contribution by MacLennan and Phillips (Chapter 9), who review all the genetic work that has shown that point mutations in the RyR1 gene are found in at least two pathological conditions (malignant hyperthermia and central core disease).

Although we know a great deal about these channels, as these chapters will show, there is still much detail to be elucidated. I wish to thank the authors for taking time out from their research activities (and eventually from their free time) to write the chapters of this book. I believe that the best reward for these efforts will be having this book become a useful source of updated information for students and scientists who are interested — or may develop an interest — in RyRs, and eventually provide the first go for further investigation on these channels.

Vincenzo Sorrentino
Milano, Italy

THE EDITOR

Vincenzo Sorrentino, M.D., is head of the Laboratory of Growth Factors and Intracellular Signaling, Department of Biology and Technology (DIBIT), Scientific Institute San Raffaele, Milano, Italy, and Associate Professor of Histology, School of Medicine, University of Siena, Italy.

Dr. Sorrentino obtained his M.D. degree from the University of Roma in 1980. After working at the Laboratory of Virology of the Istituto Superiore di Sanita in Roma from 1980 to 1983, he moved to the United States where he worked from 1983 to 1987 at the National Institutes of Health and the Sloan Kettering Cancer Center in New York. In 1988 he joined the European Molecular Biology Laboratory in Heidelberg, Germany. In 1993 he moved to his present position.

Dr. Sorrentino is a member of the American Society of Microbiology and of the European Society of Cell Biology. During the past 15 years he has published over 50 research papers mainly in the area of cellular responsiveness to growth factors. His work is supported by awards from the Public Health Minister, Telethon, Italian Association for Cancer Research, and CNR. His current research interest is the role of ryanodine receptors in intracellular signaling in excitable and non-excitable cells.

CONTRIBUTORS

Deborah L. Bennett
The Babraham Institute Laboratory of Molecular Signaling
Department of Zoology
University of Cambridge
Cambridge, United Kingdom

Michael J. Berridge
The Babraham Institute Laboratory of Molecular Signaling
Department of Zoology
University of Cambridge
Cambridge, United Kingdom

Martin D. Bootman
The Babraham Institute Laboratory of Molecular Signaling
Department of Zoology
University of Cambridge
Cambridge, United Kingdom

Timothy R. Cheek
The Babraham Institute Laboratory of Molecular Signaling
Department of Zoology
University of Cambridge
Cambridge, United Kingdom

Roberto Coronado
Department of Physiology
University of Wisconsin
School of Medicine
Madison, Wisconsin

Barbara E. Ehrlich
Laboratory of Molecular Hermeneutics
Departments of Medicine Physiology and Division of Cardiology
University of Connecticut
Farmington, Connecticut

Clara Franzini-Armstrong
Department of Cell and Developmental Biology
University of Pennsylvania
School of Medicine
Philadelphia, Pennsylvania

Antony Galione
University Department of Pharmacology
University of Oxford
Oxford, United Kingdom

Luisa Gorza
Dipartimento di Scienze Biomediche Sperimentali
Universitá di Padova
Padova, Italy

Hon Cheung Lee
Department of Physiology
University of Minnesota
Minneapolis, Minnesota

David H. MacLennan
Banting and Best Department of Medical Research
University of Toronto
C.H. Best Institute
Toronto, Ontario, Canada

Andrew R. Marks
Department of Medicine and Brookdale Center for Molecular Biology
Mount Sinai School of Medicine
New York, New York

Jeffery Morrissette
Department of Physiology
University of Wisconsin
School of Medicine
Madison, Wisconsin

Alessandra Nori
Dipartimento di Scienze
Biomediche Sperimentali
Universitá di Padova
Padova, Italy

Michael S. Phillips
Banting and Best Department of
Medical Research
University of Toronto
C.H. Best Institute
Toronto, Ontario, Canada

Vincenzo Sorrentino
DIBIT, Scientific Institute
San Raffaele, Milano, and
University of Siena
Siena, Italy

Robin Summerhill
University Department of
Pharmacology
University of Oxford
Oxford, United Kingdom

Pompeo Volpe
Istituto Pluridisciplinare
di Patologia Generale
Universitá di Messina
Messina, Italy

für Alexandra

CONTENTS

Chapter 1

ULTRASTRUCTURAL STUDIES ON FEET/ RYANODINE RECEPTORS

Clara Franzini-Armstrong

CONTENTS

1. FEET AND RYANODINE RECEPTORS

1.1. FEET ARE HIGHLY CONSERVED STRUCTURAL COMPONENTS OF MUSCLE FIBERS

Feet are large, periodically arranged structures which bridge the gap between the sarcoplasmic reticulum (SR) and either the surface membrane or the

0-8493-8543-1/95/$0.00+$.50

transverse (T) tubules at the location of junctions between the two membranes (Figures 1a and b).[1] These are the sites at which an interaction occurs between surface membrane/T tubules and SR, resulting in the transduction of depolarization into calcium release from the SR.

The SR membrane has two structurally and functionally distinct domains: one, junctional, which is occupied by feet and is in most cases associated with surface membrane/T tubules; and a second, free, occupied by the calcium pump protein or Ca ATPase.

Calsequestrin is an internal calcium binding protein of the SR which is mostly, but not exclusively, contained within the terminal region of the SR participating in junction formation. The presence of calsequestrin in the junctional region is instrumental in allowing purification of vesicles enriched in junctional domains (the so-called heavy SR fraction). From this, by further extraction of the ATPase, an almost pure fraction of junctional SR membrane can be obtained.[2] Feet are visible on the surface of heavy SR and in the purified jSR membrane, showing that the protein is an SR component.

Feet have been recognized in the SR of representatives of the entire animal kingdom, and in all varieties of muscles. Indeed feet are a highly conserved structural feature of muscle fibers, being essentially identical at low levels of resolution.

1.1.1. Structure of Feet *In Situ*

Rotary shadowing of isolated heavy SR vesicles shows the side of the feet facing the T tubules. This is a large tetramer (quatrefoil) composed of four roughly spherical subunits centered around a central depression (Figures 1c and e).[3] In grazing views of the junctional SR, both in the isolated vesicles (Figure 1c) and *in situ* (Figure 1h), feet interact in the proximity of their corners. The four subunits of the foot are barely visible in thin sections, but they are visualized after filtering and two-dimensional reconstruction in the muscle of an invertebrate.[4]

Feet are spaced at intervals of 25 to 31 nm. This means that, in thin sections cut across the junctional gap, one layer of feet is seen in the thinnest possible sections and two or more superimposed feet are present in sections of standard thickness. It is thus not surprising that variations in the appearance of feet have been described. Two images are predominant: two dense lines perpendicular to the two junctional membranes; and a single line parallel to the two membranes (Figure 1a). These appearances are consistent with different views of the quatrefoil structure: the two perpendicular lines derive from an orientation in which two sides of the foot are parallel to the plane of the section and two subunits are superimposed on each side of the central region; the parallel line may arise from an orientation in which the diagonal of the square foot is perpendicular to the plane of the section and three subunits align to form the visible line.

Images from thin sections indicate that the feet reach all the way from the SR membrane to the cytoplasmic surface of T tubules, and thus they are probably responsible for the size of the junctional gap.

The presence of other identified components of the junctions has not yet been directly detected by electron microscopy (EM), since the structure of the junctional gap is dominated by the large feet. One of the components of the triad is triadin, an intrinsic protein of the junctional SR membrane in both skeletal and cardiac muscle.[5,6] A second component is the FK-506-binding protein,[7] which is closely associated with the feet, and may have some role in the permeation properties of the channel (see Chapter 2). So far, a way of directly observing these components has not been found.

1.2. IDENTIFICATION OF FEET WITH RYANODINE RECEPTORS/SR CA RELEASE CHANNELS

Identification of the feet with the sites of calcium release from the SR has resulted from a stimulating series of converging experiments. It is clear that e-c coupling involves a communication between T tubules and SR and that this occurs at their junction. Thus, membrane fractions containing either heavy SR or triads have become the focus of many laboratories. The first hint came from the observation that a high molecular weight component of the triad exchanged its ^{125}I label with T tubules and thus was a likely candidate for being a component of feet. This led to the first isolation and characterization of a component of the feet, with a molecular weight of approximately 350,000.[8] At this point, two substances which affect calcium release from the SR, doxorubicin and ryanodine, became available in radioactive form and were shown to have high affinity for junctional SR and some high molecular weight component in it.[9-11] The site of calcium release from the SR was identified to be a channel present in the heavy SR fractions, where feet are located.[12] The final preparatory step was the finding that isolation of a "ryanodine receptor" (or RyR), i.e., a molecule which maintained high affinity for ryanodine, could be achieved by using a mild detergent, and that this resulted in the solubilization of a very high molecular weight (1.2×10^6) RyR.[10] Purification of the RyR and its identification with the junctional feet quickly followed in three laboratories,[13-15] and it was demonstrated that the reconstituted molecule had the same properties as the calcium release channel of the heavy SR.[14] Properties of the isolated and *in situ* RyRs have recently been reviewed.[16,17] The purified RyR is a homotetramer with a 30 S sedimentation coefficient, and is composed of four large chains of approximately 500 kDa each (more precisely 565 kDa from the sequence,[18] see Chapter 6). Identification of the RyR/Ca release channel within the feet is immediate, due to the unusually large size and shape of the molecule (compare Figures 1c and e with 1d through g).

In the rest of the chapter the terms RyR, SR Ca release channel, and foot are used to indicate the molecule identified chemically, electrophysiologically, and by EM.

Immunolabeling for RyRs in skeletal and cardiac muscle confirms the location of RyRs in sites where feet are seen by EM (see Reference 19 for a review). Homotetrameric RyRs with similar, but not identical, molecular weights

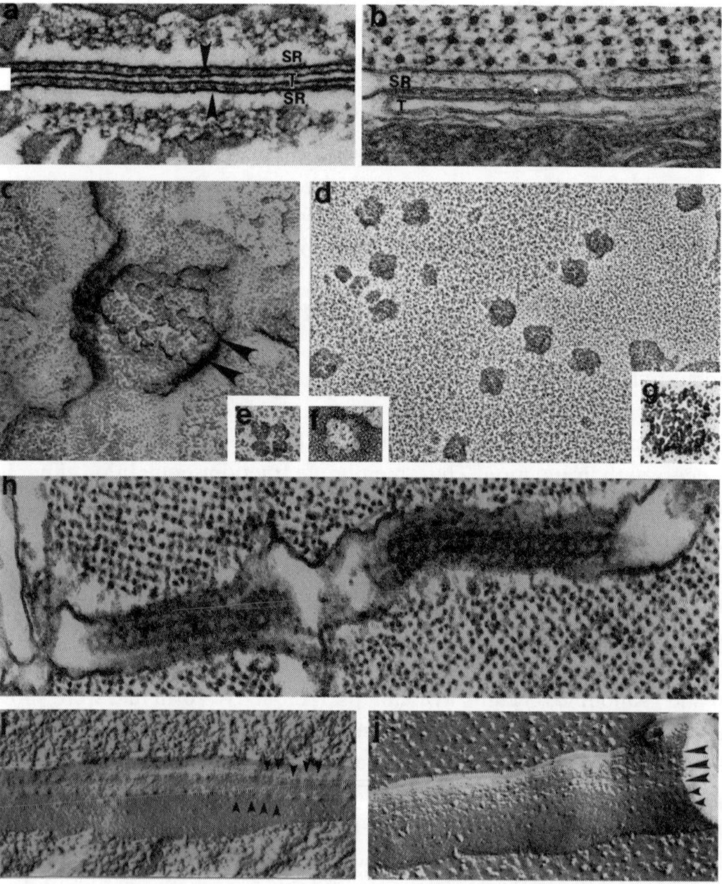

FIGURE 1. (**a** and **b**) Thin sections through T-SR junctions in the toadfish swimbladder (**a**) and the dragonfly flight muscle (**b**). The gap between T tubule (T) and sarcoplasmic reticulum (SR) is filled with feet. Most of the feet (**a**) appear as lines bisecting the gap, others (arrowhead) more clearly span the entire gap, joining SR to T tubules × 117,000 and 130,000). (**c** and **e**) Isolated heavy SR vesicles have feet associated with their surfaces. In (**c**), from guinea pig,[3] feet maintain the native disposition in two rows (arrowheads). In (**e**), from crayfish,[4] the single foot shows four equal subunits and a central depression [(**c**) × 190,000; (**e**) × 286,000]. (**d, f,** and **g**) Purified RyRs from rabbit skeletal muscle,[15] rotary shadowed (**d** and **f**) and negatively stained (**g**). The isolated molecule is composed of a larger foot region and a smaller, central raised platform, with four subunits rotated by 45° relative to the foot region (see References 14 and 23). [(**d**) × 174,000; (**f**) × 286,000; (**g**), × 482,000]. (**h**) Grazing view of the junctional gap in the triad of a fish muscle.[1] Feet are disposed in a tetragonal array, forming two or occasionally three rows. Feet are positioned diagonally, but with a slight skew relative to the long axis of the T tubule (see Reference 3); (× 93,000). (**i**) Freeze-fracture through the junctional domain of the SR in toadfish swimbladder muscle.[15] The cytoplasmic leaflet is occupied by small bumps with a spacing equal to that of the feet, corresponding to the channel region of the RyR. In some of these (arrowheads), the four-subunit structure of the

and properties have been identified in a large variety of skeletal muscles, in the cardiac and smooth muscle of vertebrates, and in striated muscles of invertebrates.[4,17] It is thus fair to assume that the feet visible by EM in all types of muscle fibers represent the same molecular species.

The converse is not true: EM may miss detection of feet where RyRs are not arranged in arrays. In skeletal muscle, extrajunctional RyRs have been detected by immunocytochemistry, but not directly visualized.[20] In addition, despite the fact that RyRs have been identified in a variety of nonmuscle cells[21] (see Chapter 6), arrays of feet have been detected only in Purkinje cells.[22]

1.2.1. Substructure of the Isolated Channel

The isolated RyR/Ca release channel has two components: a large tetrameric assembly which clearly corresponds to the foot and represents the cytoplasmic, hydrophylic domain of the molecule, and a smaller domain, rotated by 45°, which protrudes from one side of the molecule and constitutes the intramembrane domain (Figures 1d through g).[14,15,23] 3-D reconstructions indicate that this domain is sufficiently large to span the thickness of the SR membrane.[23] Freeze-fracture of the junctional SR membrane shows intramembrane domains of the foot protein (Figure li)[15,24] which are consistent with the size of the raised domain in the isolated molecule.[23]

In detailed three-dimensional reconstructions from isolated molecules,[23] both cytoplasmic and membrane domains are composed of four equal components, constituting a fourfold symmetric molecule. This indicates that each of the four polypeptides contributes equally to the cytoplasmic and membrane domains of the molecule.

The most striking aspect of the transmembrane assembly is a fairly large (2 to 3 nm diameter) central channel which seems to have a central plug near the external edge of the membrane. This might be the pathway for calcium permeation. However, since the release channel is capable of four conducting substates (see Chapter 2) it is possible that a channel is located in each subunit.

The cytoplasmic assembly is a highly hydrated structure with numerous cavities and channels between ten identifiable domains, linking to the intramembrane segment on one side, and forming four large lobes with a fairly flat top at the other. The lobes give a square appearance to the feet, and are the sites that reach toward the T tubules. The pathway for calcium diffusion through this structure is quite open, and the structure appears like a complex trellis designed for structural strength, but allowing diffusion of solutes. Sites

---

channel is seen (× 105,000). (**j**) Freeze-fracture of the junctional SR from the triad of a frog sartorius. The junctional membrane is occupied by three rows of RyR channels (large arrowheads). However, two additional rows of similar structures (small arrowheads) are also visible. These are on the side of the junctional SR cisternae, where feet are not usually located (× 116,000).

of interaction of calmodulin with the RyR have been mapped to the four corners of the foot.[25] It is interesting that these sites are at some distance from the channel portion of the molecules, which is regulated by calmodulin.

The best estimates of the overall dimensions of the cytoplasmic assemblies (feet) in the x-y direction is 29 × 29 nm.[23] The spacing of feet in arrays (see model in Reference 26) on the other hand, would indicate an effective size of 27 nm or less. This probably means that some overlap and/or interdigitating of the corners of the molecule occur in the formation of arrays. The z-axis dimension of the cytoplasmic assembly (12 nm) is consistent with a molecule that crosses the entire junctional gap.

2. DISPOSITION OF FEET *IN SITU*

2.1. FEET ARE DISPOSED IN ORDERLY ARRAYS

In all muscle cells, feet are grouped in assemblies which are easily recognizable in thin sections and are detectable as intense foci when RyRs are immunolabeled with fluorescent antibodies for the light microscope.[1,19] The junctional surface of the SR is fully covered by feet assemblies. The elongated junctional SR domains of twitch fibers of vertebrates have two to three parallel rows of feet on their surfaces (Figure 1h), and the wide junctional cisternae of slow fibers, cardiac muscle, and muscles of invertebrates (Figure 2a) have extended, multiple rows. In all cases, the arrangement is tetrameric and involves a specific interaction between corners of adjacent feet and probably some interdigitation between the molecules.

In skeletal muscle, all feet have equal orientation. The array is quite closely mimicked by representing the feet with four equal spheres, and locating them so that the centers of subunits of adjacent feet follow along the same line (Figure 2b).[3] This results in some overlap between the subunits of adjacent feet and in a skew of the feet relative to the axis of symmetry of the overall array. The predicted interaction between the feet is near the corners, consistent with the structure of the reconstructed molecule[23] which would allow a close fit between adjacent feet in that region.

In muscles of some arthropods, the arrangement is dimeric (Figure 2c):[4] each unit cell is composed of two feet displaced along the diagonal of the cell and tilted, one in relation to the other, by 23°. A schematic representation of the two dispositions observed in vertebrates and invertebrates (Figures 2b and c) shows that while interaction in the arrays involves similar regions of the molecule, details of their association must be quite different. A detailed image of invertebrate feet is now needed to understand this difference.

The precise disposition of feet in cardiac muscle of vertebrates has not been defined, due to the inherent difficulty of obtaining grazing views of the junctions in these muscles.

In peripheral couplings of developing skeletal muscle, the disposition of feet has not been directly observed, but it may be inferred to be similar to that of adult muscle from the observation that tetrads, which are associated with the feet (see below), have the same disposition as in the adult triad.

2.1.1. Formation of Arrays is an Intrinsic Property of the RyR

The following evidence indicates that assembly into tetragonal arrays is an intrinsic property of the RyR:

1. Arrays of feet with the same spacing as in muscle are formed when RyRs are expressed in Chinese hamster ovary cells (Figure 2g);[27]
2. In muscles of invertebrates, arrays of feet extend beyond T tubules;[4]
3. In cardiac muscle, arrays of feet are formed in SR domains which do not face either the surface membrane or T tubule;[28]
4. Arrays of feet remain associated with the junctional SR membrane after it has been detached from the T tubules.

Since the cytoplasmic domains of RyRs apparently touch each other, while the intramembrane domains, being smaller, are separated by some distance (see below), formation of arrays is likely to be a property of the cytoplasmic domains.

2.1.2. Truncated RyRs in Skeletal Muscle?

The intramembrane domains of RyRs, forming the Ca release channels, are visible in freeze-fracture replicas as barely detectable quatrefoil structures with a central depression (Figure 1i).[15,24] Two rows of such structures occupy the junctional SR membrane underlying the feet. In some cases, however, the two junctional rows of RyRs channels are flanked on either side by two or more additional rows of similar intramembrane structures (Figure 1j). These additional rows are located in the nonjunctional region of the SR, where no feet are visible. This suggests the presence of a truncated RyRs, of the type predicted from the short sequence of brain tissue.[29]

2.2. INTERACTIONS BETWEEN FEET AND SURFACE MEMBRANES

2.2.1. Skeletal Muscle

In skeletal muscle, all arrays of feet are seen either at sites of T-SR or at surface membrane-SR junctions. T-SR junctions (triads and dyads) are present in all adult muscle fibers, with the exception of those that do not have T tubules (amphioxus is actually the only known example). Surface membrane-SR junctions (peripheral couplings) are present at early developmental stages, before T tubules form, but usually decline as the T system takes over and are rare in most adult fibers. Some adult fibers, such as the slow tonic fibers in the frog, have a mixture of internal and peripheral junctions.

Some RyRs not forming arrays may be present at some distance from the junctional SR.[20] However, the classical local stimulation experiments in frog fibers show that the contractile response to depolarization of a portion of a

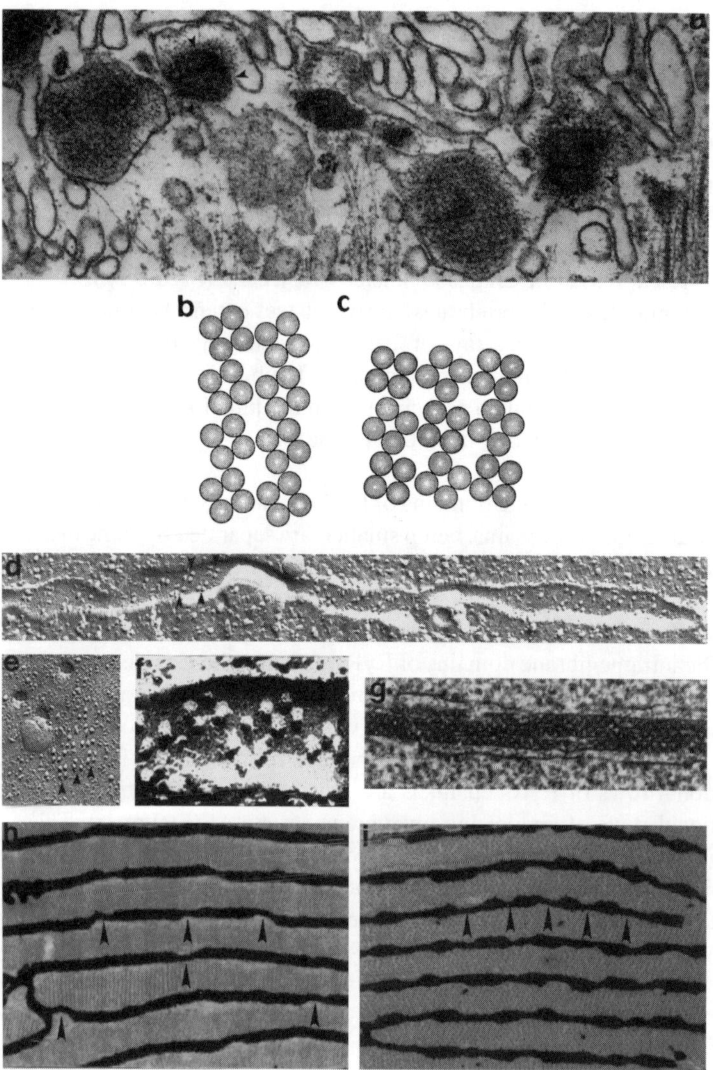

FIGURE 2. (**a**) Grazing view of dyads in the muscle of a scorpion.[4] Extensive arrays of tetragonally disposed feet occupy the junctional gap. The darker areas (arrowheads) are profiles of T tubules. Note that feet extend beyond the area covered by T tubules (× 66,000). (**b** and **c**) Comparison of arrays of feet in vertebrates (**b**) and some invertebrates (**c**).[4] In the latter the arrangement is dimeric. (**d** through **f**) Tetrads (some indicated by arrowheads) occupy the junctional domains of T tubules (**d** and **f**), from toadfish[15] and of the plasmalemma (**e**) from a frog tonic fiber.[60] In T tubules (**d** and **f**), tetrads form two rows parallel to the T tubule's long axis. The tetrads have the same slightly skewed position as feet, indicating superimposition of the four components of the tetrads with the four subunits of feet. However, the spacing between tetrads is twice that between feet, indicating interaction of tetrads with alternate feet. This is also the case in peripheral

single T tubule network remains localized in the proximity of the T tubule. This indicates that feet located at some distance from the junctions may not play a significant role in e-c coupling of frog muscle. Imaging of the early calcium signal following stimulation of frog fibers suggests that a small component of release may occur at some distance from the junctions.[30] However, the contribution of out-of-focus intensity to the image would have to be eliminated before a quantitative estimate of the possible contribution of nonjunctional feet to e-c coupling can be fully assessed.

2.2.2. Cardiac Muscle and Muscles of Invertebrates

In cardiac muscle, feet participate in the formation of peripheral couplings and internal junctions with T tubules, just as in skeletal muscle. However, in addition to these junctions there are small (corbular SR) and large (extended junctional SR) areas of SR covered by feet arrays but not associated with either T tubules or surface membranes and located at distances of up to several microns from them.[28,31] These nonjunctional feet are identified as RyRs[19] and they are likely to contribute significantly to calcium release, since in some cases they constitute a high proportion of the total RyR content.

In muscles of invertebrates, the situation is somewhat similar to that in cardiac muscle, in the sense that arrays of feet are often found in areas not directly facing a T tubule.[4,19] In this case, however, the extrajunctional patches of feet are immediately adjacent to those participating in junctions.

3. INTERACTIONS BETWEEN FEET AND DIHYDROPYRIDINE RECEPTORS

3.1. TWO CATEGORIES OF FEET IN SKELETAL MUSCLE

3.1.1. Tetrads and Their Relationship to Feet

Tetrads are groups of four intramembrane particles visible in freeze-fracture images of junctional domains of either T tubules or surface membranes (Figures 2d through f). Tetrads are located in arrays directly related to those of the feet,[15,26,32] and their position and spacing indicate that the four components of the tetrad sit above the four subunits of a foot. The domains of the foot most likely to make contact with components of the T tubules are positioned approximately as expected from the position of the four components of the tetrads.[23]

---

couplings (**e**),[26] (**d**) × 103,000; (**e**) × 120,00; (**f**) × 330,000. (**g**) Extensive crystalline arrays of feet cover the surface of endoplasmic reticulum in CHO cells expressing the RyR.[27] The spacing is the same as in skeletal muscle, indicating ability of feet to self assemble. In CHO, cell feet do not form junctions with the surface membrane (× 80,000. (**h** and **i**) Cross sections of two muscle fibers from the swimbladder of a female (**h**) and a male (**i**) toadfish.[54] T tubules are infiltrated by the Golgi stain and their lumens are electron opaque. Short, narrow T tubule segments (some indicated by arrowheads) separate wider, junctional T tubule segments. Note that the lengths of the junctional segments and the sizes of the e-c units are different in the two muscle fibers (× 24,480).

Extensive arrays of tetrads have been seen only in a limited number of cases, including examples from fish (Figures 2d and f), amphibia (Figure 2e), and birds and mammals.[19] For the moment, there is no reason to suspect that failure to observe arrays of tetrads in other muscle is due to an absence of tetrads. Rather it may be due to the technical difficulties in their preservation and/or visualization.

An unexpected finding is that only alternate feet interact with tetrads. The direct result of this observation is that the junctional SR of skeletal muscle contains two categories of feet: one associated with tetrads, the other not. One simple explanation for this finding is that a tetrad may occupy a portion of membrane larger than a foot, possibly because its four components are located somewhat peripherally relative to the center of the foot. Feet immediately adjacent to those occupied by a tetrad would thus be hindered from interacting. The precisely alternate disposition of tetrads might require that arrays of feet and tetrads assemble simultaneously. Indeed, during development both tetrads and feet appear very early[33] and so does apposition of RyR and dihydropyridine receptors,[34] the probable components of tetrads.

3.1.2. Are Tetrads Composed of Dihydropyridine Receptors?

Dihydropyridine receptors (DHPR) are the slowly activating calcium channels of striated muscle fibers which are thought to act as voltage sensors for excitation-contraction coupling.[35,36] It has been proposed that tetrads consist of four DHPRs.[15] Several lines of evidence confirm this hypothesis. First, DHPRs, as tetrads, occupy the junctional domains of surface membrane/T tubules.[37,38] Secondly, tetrads are missing from muscle fibers of dysgenic mice,[32] in which a mutation of the DHPR results in lack of the protein and block of e-c coupling, despite the fact that junctions and assemblies of feet are present. Finally, transfection of cultured dysgenic fibers with cDNA for the DHPR restores tetrads.[26]

The position of the tetrads provides strong evidence for the formation of a large molecular complex spanning the entire junction, from the lumen of the T tubule to the lumen of the SR, and maintaining a fourfold symmetry throughout. This complex would be most appropriate for allowing a direct molecular interaction between the voltage sensor of the e-c coupling and the SR calcium release channel, as has been proposed by Schneider and Chandler.[39]

Additional support for such a scheme comes from observations that the cytoplasmic loop between repeats II and III of DHPR confers skeletal-type e-c coupling properties to the cardiac DHPR,[36] and the same loop has a direct effect on the function of the RyR.[40] However, despite copurification of DHPRs and RyRs, direct linking of the two proteins has not been demonstrated, leading to the hypothesis that some other component of the junction, namely triadin, may intervene.[41,42] Triadin associates with RyRs and DHPRs, but does not need the latter for its presence in the junctions.[43] However, there is disagreement on the role of triadin in the junctions.[41,44]

Natural and induced null mutations for the DHPR[45] and for the RyR[46] have unexpectedly shown that T-SR junctions (triads) form in the absence of DHPRs/tetrads[32,43] and of RyRs/feet.[47] In the latter case, tetrads are also absent. Interestingly, functional expression of DHPRs and RyRs in CHO cells does not result either in the formation of junctions or of tetrads, although the feet do assemble into ordered arrays.[27] These results imply the need of components other than RyRs and DHPRs in the formation of the junction and indicate that interaction, directly or indirectly, between DHPRs and RyRs is needed for the formation of tetrads.

Interesting functional possibilities of the alternate relationship between tetrads and feet have been presented.[35] It is important to note that the possibility of two functional types of feet is not linked to the presence of two isoforms of the RyRs, since the toadfish swimbladder, which has a single isoform,[48] is one of the best-described examples of alternate disposition of tetrads.[15]

If tetrads are composed of four DHPRs, the DHPR/RyR ratio predicted from the structure is 2:1. At the moment, some biochemical information does not agree with this expectation. High affinity binding of tritiated agonists for the two molecules indicate a considerably lower ratio.[16] Extrajunctional RyRs would result in a decreased ratio, but it is not clear that the density of extrajunctional RyRs is sufficiently high to account for the biochemical data. Clearly, this discrepancy will need to be resolved before full credit is given to the predictions based on structural considerations.

3.2. CARDIAC MUSCLE AND MUSCLES OF INVERTEBRATES

Cardiac muscles and muscles of arthropods have different e-c coupling properties from skeletal muscle. While the latter can be stimulated to contract in the absence of extracellular calcium and of calcium currents, the two former types of muscle have calcium currents that precede tension, and do not release internal calcium unless calcium permeates through the DHPRs.[49,50]

In cardiac muscle, DHPRs are located in intense foci related to the position of the junctional SR.[51,52] Grouping of DHPRs at junctional sites is also confirmed by the observation that most DHPRs can be recovered with the dyadic fraction in isolated membranes from rat ventricular muscle.[6] Freeze-fracture of cardiac muscle, however, does not reveal the presence of tetrads; rather, it shows presumptive DHPRs in an apparently random position within the junctional domains of the surface membrane.[52] This indicates a substantial difference from skeletal muscle and suggests that interaction between DHPRs and RyRs may not be direct in cardiac muscle.

Less is known about muscles of invertebrates, but a DHP binding protein has been characterized in muscles of crayfish.[53] Freeze-fracture shows that in all muscles of arthropods tetrads are missing but, as in cardiac muscle, randomly distributed intramembrane particles probably representing DHPRs are located in T tubules and surface membrane at the junctions.[1]

Thus in cardiac and invertebrate muscles, DHPRs and RyRs are sufficiently near each other at the junctions that DHPR activation may fairly directly affect the RyRs, but a direct interaction between the two molecules may not be possible.

4. E-C COUPLING UNITS

In images where the T tubules are delineated by some electron dense marker, it is clear that T tubules have two different shapes: small tubules with round cross sections alternate with dilated tubules with two flat surfaces (Figures 2h and i).[1] The latter form junctions with the SR, while the round tubules provide spacers between the junctional sites. We have named the discrete junctional sites Ca release units[33] and e-c units,[19] referring to their function as sites of Ca^{2+} release and of interaction between surface membrane/T tubules and SR. By modulating the frequency and size of the e-c units along the T tubules the total content of feet/RyRs can be fitted to the fiber requirement, without large alterations in the content and distribution of T tubules. In guinea pig leg muscles, the content of feet is the most precise discriminatory component between different fiber types.[1] The size and spacing of the individual units varies considerably from one muscle fiber to another. In toadfish, the swimbladder muscle of males (Figure 2i) has units of approximately 6 to 14 feet, while in females (Figure 2h) the units are longer and may have up to 40 feet.[54] In slow tonic fibers the units are placed at long intervals along the T tubules. These variations may have functional significance, beyond modulating feet content.[1] A possibility comes from the observation that the feet within a unit have a better chance of interacting with each other than with feet of other units. If initial activation of one or more feet in a unit resulted in facilitation of the other feet within the same unit, then the size of the units would have a direct impact on the size of the calcium transient and on its rate of rise. Modulating the size of the e-c units thus would be a convenient way of tailoring amplification of the effect of depolarization to fit fiber requirements. In this scheme, the steep voltage dependence of activation in skeletal muscle would apply to the activation of e-c units rather than of single channels. Separate, depolarization-dependent activation of each unit would also be consistent with the results of the local stimulation experiments, in which calcium release is strictly dependent on T tubule depolarization and never spreads on its own. Distance between the units would affect e-c coupling only if the facilitation mentioned above was due to some diffusible agent, say calcium, rather than to a direct molecular interaction between the feet.

Cardiac muscle and muscles of invertebrates are more complex than skeletal muscle. On the one hand, calcium release from the SR seems to depend on sarcolemmal and T tubular I_{Ca}, thus requiring that it is the result of events at the surface membrane; on the other hand, release from the internal junctional SR not associated with T tubules must be activated indirectly, by some regenerative mechanism.[28,31] Models based on calcium-activated calcium release result in three apparent paradoxes:

1. Calcium release is graded according to calcium entry,[55,56] indicating that calcium release, once initiated, does not normally spread as a wave through the fiber;
2. On the other hand, participation of corbular and extended junctional SR in e-c coupling of cardiac muscle requires spread of activity from one coupling to the next;
3. Calcium entering the fiber through the DHP sensitive channels seems to be more effective than that liberated from the SR in activating release of cardiac and invertebrate muscle fibers.[50,57]

Clearly, either calcium is not the activating factor[58] or some quite complex control mechanisms are needed to fit all the observed phenomena.[59] Comparative morphology suggests that e-c coupling in cardiac muscle varies considerably between fibers that have large amounts of extended junctional SR and those in which most of the junctional SR does form junctions. Functional comparison between extreme examples of this variability may be quite fruitful.

REFERENCES

1. **Franzini-Armstrong, C.,** The sarcoplasmic reticulum and the transverse tubules, in *Myology,* Engel, A.G. and Franzini-Armstrong, C., Eds., 2nd ed., McGraw-Hill, New York, 1994, 176.
2. **Fleischer, S. and Inui, M.,** Biochemistry and biophysics of excitation-contraction coupling, *Annu. Rev. Biophys. Biophys. Chem.,* 18, 333, 1989.
3. **Ferguson, D. G., Schwartz, H. W., and Franzini-Armstrong, C.,** Subunit structure of junctional feet in triads of skeletal muscle: a freeze-drying, rotary-shadowing study, *J. Cell Biol.,* 99, 1735, 1984.
4. **Loesser, K. E., Castellani, L., and Franzini-Armstrong, C.,** Disposition of junctional feet in muscles of invertebrates, *J. Muscle Res. Cell Motil.,* 13, 161, 1992.
5. **Kim, K. C., Caswell, A. H., Talvenheimo, J. A., and Brandt, N. R.,** Isolation of a terminal cisterna protein which may link the dihydropyridine receptor to the junctional foot protein in skeletal muscle, *Biochemistry,* 29, 9283, 1990.
6. **Brandt, N. R., Caswell, A. H., Carl, S. A., Ferguson, D. G., Brandt, T., Brunschwig, J. P., and Bassett, A. L.,** Detection and localization of triadin in rat ventricular muscle, *J. Membr. Biol.,* 131, 219, 1993.
7. **Timerman, A. P., Ogunbumni, E., Freund, E., Wiederrecht, G., Marks, A. R., and Fleischer, S.,** The calcium release channel of sarcoplasmic reticulum is modulated by FK-506-binding protein. Dissociation and reconstitution of FKBP-12 to the calcium release channel of skeletal muscle sarcoplasmic reticulum, *J. Biol. Chem.,* 268, 22992, 1993.
8. **Kawamoto, R. M., Brunschwig, J. P., and Caswell, A. H.,** Localization by immunoelectron microscopy of spanning protein of triad junction in terminal cisternae/triad vesicles, *J. Muscle Res. Cell Motil.,* 9, 334, 1988.

9. **Fleischer, S., Ogunbunmi, E. M., Dixon, M. C., and Fleer, A. M.,** Localization of Ca^{2+} release channels with ryanodine in fast skeletal muscle, *Proc. Natl. Acad. Sci. U.S.A.,* 82, 7256, 1985.
10. **Pessah, I. N., Francini, A. O., Scales, D. J., Waterhouse, A. L., and Cassida, J. E.,** Calcium-ryanodine receptor complex. Solubilization and partial characterization from skeletal muscle junctional sarcoplasmic reticulum vesicles, *J. Cell Biol.,* 61, 8643, 1986.
11. **Zorzato, F., Margreth, A., and Volpe, P.,** Direct photoaffinity labeling of junctional sarcoplasmic reticulum with [^{14}C]doxorubicin, *J. Biol. Chem.,* 261, 13252, 1986.
12. **Smith, J. S., Coronado, R., and Meissner, G.,** Sarcoplasmic reticulum contains adenine nucleotide activated calcium channels, *Nature,* 316, 446, 1985.
13. **Inui, M., Saito, A., and Fleischer, S.,** Purification of the ryanodine receptor and identity with feet structures of junctional terminal cisternae of sarcoplasmic reticulum from fast skeletal muscle, *J. Biol. Chem.,* 262, 1740, 1987.
14. **Lai, F. A., Erickson, H. P., Rousseau, E., Liu, Q. Y., and Meissner, G.,** Purification and reconstitution of the calcium release channel from skeletal muscle, *Nature,* 331, 315, 1988.
15. **Block, B. A., Leung, A., Campbell, K. P., and Franzini-Armstrong, C.,** Structural evidence for direct interaction between the molecular components of the transverse tubules/sarcoplasmic reticulum junction in skeletal muscle, *J. Cell Biol.,* 107, 2587, 1988.
16. **Meissner, G.,** Ryanodine receptor/Ca^{2+} release channels and their regulation by endogenous effectors, *Annu. Rev. Physiol.,* 56, 485, 1994.
17. **Coronado, R., Morrissette, J., Sukhareva, M., and Vaughan, D. M.,** Structure and function of ryanodine receptors, *Am. J. Physiol.,* 266, C1485, 1994.
18. **Takeshima, H., Nishimura, S., Matsumoto, T., Ishida, H., Kaugawa, K., Minamino, N., Matsuo, H., Ueda, M., Hanaoka, M., Hirose, T., and Numa, S.,** Primary structure and expression from complementary DNA of skeletal muscle ryanodine receptor, *Nature,* 339, 439, 1989.
19. **Franzini-Armstrong, C. and Jorgensen, A. O.,** Structure and development of e-c coupling units in skeletal muscle, *Annu. Rev. Physiol.,* 56, 509, 1994.
20. **Dulhunty, A. F., Junankar, P. R., and Stanhope, C.,** Extrajunctional ryanodine receptors in the terminal cisternae of mammalian skeletal muscle fibres, *Proc. R. Soc. London Ser. B,* 247, 69, 1992.
21. **Sorrentino, V. and Volpe, P.,** Ryanodine receptors: how many, where and why?, *Trends Pharmacol. Sci.,* 14, 98, 1993.
22. **Henkart, M., Landis, D. M. D., and Reese, T. S.,** Similarity of junctions between plasma membrane and endoplasmic reticulum in muscle and neurones, *J. Cell Biol.,* 70, 338, 1976.
23. **Radermacher, M., Rao, V., Grassucci, R., Frank, J., Timerman, A. P., Fleischer, S., and Wagenknecht, T.,** Cryo-electron microscopy and three-dimensional reconstruction of the calcium release channel/ryanodine receptor from skeletal muscle, *J. Cell Biol.,* 127, 411, 1994.
24. **Kelly, D. E. and Kuda, A. M.,** Subunits of the triadic junction in skeletal muscle as revealed by freeze-fracture, *J. Ultrastruct. Res.,* 68, 220, 1979.
25. **Wagenknecht, T., Berkowitz, J., and Grassucci, R.,** Localization of calmodulin binding sites on the ryanodine receptor from skeletal muscle by electron microscopy, *Biophys. J.,* 67, 2286, 1994.
26. **Takekura, H., Bennet, L., Tanabe, T., Beam, K. G., and Franzini-Armstrong, C.,** Restoration of junctional tetrads in dysgenic myotubes by dihydropyridine receptor cDNA, *Biophys. J.,* 67, 793, 1994.
27. **Takekura, H., Takeshima, H., Nishimura, S., Takahashi, M., Tanabe, T., Flockerzi, V., Hoffman, F., and Franzini-Armstrong, C.,** Co-expression in CHO cells of two muscle proteins involved in e-c coupling, *J. Mus. Res. Cell Mot.,* accepted for publication, 1995.
28. **Sommer, J. R., Bossen, E., Dalen, H., Dolber, P., High, T., Jewett, P., Johnson, E. A., Junker, J., Leonard, S., Nassar, R., Scherer, B., Spach, M., Spray, T., Taylor, I., Wallace, N. R., and Waugh, R.,** To excite a heart: a bird's view, *Acta Physiol. Scand.,* 142 (Suppl. 599), 5, 1991.

29. **Hakamata, Y., Nakai, J., Takeshima, H., and Imoto, K.,** Primary structure and distribution of a novel ryanodine receptor/calcium release channel from rabbit brain, *FEBS Lett.,* 312, 229, 1992.
30. **Escobar, A., Monck, J., Fernandez, J., and Vergara, J.,** Localization of the site of Ca^{2+} release at the level of a single sarcomere in skeletal muscle fibres, *Nature,* 367, 739, 1994.
31. **Bossen, E. H., Sommer, J. R., and Waugh, R. A.,** Comparative stereology of the mouse and finch left ventricle, *Tissue Cell,* 10, 773, 1978.
32. **Franzini-Armstrong, C., Pincon-Raymond, M., and Rieger, F.,** Muscle fibers from dysgenic mouse *in vivo* lack a surface component of peripheral couplings, *Dev. Biol.,* 146, 364, 1991.
33. **Takekura, H., Sun, X.-H., and Franzini-Armstrong, C.,** Development of the excitation-contraction coupling apparatus in skeletal muscle. Peripheral and internal calcium release units are formed sequentially, *J. Muscle Res. Cell Motil.,* 15, 102, 1994.
34. **Yuan, S., Arnold, W., and Jorgensen, A. O.,** Biogenesis of transverse tubules and triads: immunolocalization of the 1,4-dihydropyridine receptor, TS28, and the ryanodine receptor in rabbit skeletal muscle developing *in situ, J. Cell Biol.,* 112, 289, 1991.
35. **Rios, E., Ma, J., and Gonzalez, A.,** The mechanical hypothesis of excitation-contraction (EC) coupling in skeletal muscle, *J. Muscle Res. Cell Motil.,* 12, 127, 1991.
36. **Tanabe, T., Beam, K. G., Adams, B. A., Nicodome, T., and Numa, S.,** Regions of the skeletal muscle dihydropyridine receptor critical for excitation-contraction coupling, *Nature,* 346, 567, 1990.
37. **Jorgensen, A. O., Shen, A. C.-Y., Arnold, W., Leung, A. T., and Campbell, K. P.,** Subcellular distribution of the 1,4-Dihydropyridine receptor in rabbit skeletal muscle *in situ.* An immunofluorescence and immunocolloidal gold-labeling study, *J. Cell Biol.,* 109, 135, 1989.
38. **Flucher, B. E., Morton, M. E., Froehner, S. C., and Daniels, M. P.,** Localization of the $alpha_1$ and $alpha_2$ subunits of the dihydropyridine receptor and ankyrin in skeletal muscle triads, *Neuron,* 5, 339, 1990.
39. **Schneider, M. F. and Chandler, W. K.,** Voltage dependent charge movement in skeletal muscle: a possible step in excitation-contraction coupling, *Nature,* 242, 244, 1973.
40. **Lu, X., Xu, L., and Meissner, G.,** Activation of the skeletal muscle calcium release channel by a cytoplasmic loop of the dihydropyridine receptor, *J. Biol. Chem.,* 269, 6511, 1994.
41. **Brandt, N. R., Caswell, A. H., Wen, S.-R., and Talvenheimo, J. A.,** Molecular interactions of the junctional foot protein and dihydropyridine receptor in skeletal muscle triads, *J. Membr. Biol.,* 113, 237, 1990.
42. **Caswell, A. H., Brandt, N. R., Brunschwig, J. P., and Purkerson, S.,** Localization and partial characterization of the oligomeric disulfide-linked molecular weight 95,000 protein (triadin) which binds the ryanodine and dihydropyridine receptors in skeletal muscle triadic vesicles, *Biochemistry,* 30, 7507, 1991.
43. **Flucher, B. E., Andrews, S. B., Fleischer, S., Marks, A. R., Caswell, A., and Powell J.A.,** Triad formation: organization and function of the sarcoplasmic reticulum calcium release channel and triadin in normal and dysgenic myotubes, *J. Cell Biol.,* 123, 1161, 1993.
44. **Knudson, C. M., Stang, K. K., Jorgensen, A. O., and Campbell, K. P.,** Biochemical characterization and ultrastructural localization of a major junctional sarcoplasmic reticulum glycoprotein (triadin), *J. Biol. Chem.,* 268, 12637, 1993.
45. **Chaudari, N.,** A single nucleotide deletion in the skeletal muscle specific calcium channel transcript of muscular dysgenesis (mdg) mice, *J. Biol. Chem.,* 267, 25636, 1992.
46. **Takeshima, H., Iino, M., Takekura, H., Nishi, M., Kuno, J., Minowa, O., Takano, H., and Noda, T.,** Excitation-contraction uncoupling and muscular degeneration in mice lacking junctional skeletal muscle ryanodine-receptor gene, *Nature,* 369, 556, 1994.
47. **Takekura, H., Nishi, M., Noda, T., Takeshima, H., and Franzini-Armstrong, C.,** Abnormal junctions between surface membrane and sarcoplasmic reticulum in skeletal muscle with a mutation targeted to the ryanodine receptor, *Proc. Natl. Acad. Sci. U.S.A.,* 92, 3381, 1995.

48. **O'Brien, J., Meissner, G., and Block, B. A.,** The fastest contracting muscles of nonmammalian vertebrates express only one isoform of the ryanodine receptor, *Biophys. J.,* 65, 2418, 1993.
49. **Nabauer, M., Gallewaert, G., Cleemann, L., and Morad, M.,** Regulation of calcium release is gated by calcium current, not gating charge, in cardiac myocytes, *Science,* 244, 800, 1989.
50. **Gyorke, S. and Palade, P.,** Role of local Ca^{2+} domains in activation of Ca^{2+}-induced Ca^{2+} release in crayfish muscle fibers, *Am. J. Physiol.,* 264, C1505, 1993.
51. **Carl, S. L., Felix, K., Caswell, A. H., Brandt, N. R., Ball, W. J., Vaghy, P. L., Meissner, G., and Ferguson, D. G.,** Immuno localization of sarcolemmal dihydropyridine receptor and sarcoplasmic reticular triadin and ryanodine receptor in rabbit ventricle and atrium, *J. Cell Biol.,* 129, 673, 1995.
52. **Sun, X.-H., Protasi, F., Takahashi, M., Takeshima, H., Ferguson, D. G., and Franzini-Armstrong, C.,** Molecular architecture of membranes involved in excitation-contraction coupling of cardiac muscle, *J. Cell Biol.,* 129, 659, 1995.
53. **Krizanova, O., Novotova, M., and Zachar, J.,** Characterization of DHP binding protein in crayfish striated muscle, *FEBS Lett.,* 267, 311, 1990.
54. **Appelt, D., Shen, V., and Franzini-Armstrong, C.,** Quantitation of Ca ATPase, feet and mitochondria in super fast muscle fibres from the toadfish, *Opsanus tau, J. Muscle Res. Cell Motil.,* 12, 543, 1991.
55. **Cannell, M. B., Berlin, J. R., and Lederer, W. J.,** Effect of membrane potential changes on the calcium transient in single rat cardiac muscle cells, *Science,* 238, 1419, 1987.
56. **Cleeman, L. and Morad, M.,** Role of Ca^{2+} channels in cardiac excitation-contraction coupling in the rat: evidence from Ca^{2+} transients and contraction, *J. Physiol.,* 432, 283, 1991.
57. **Stern, M. D. and Lakatta, E. G.,** Excitation-contraction coupling in the heart—the state of the question, *FASEB J.,* 6, 3092, 1992.
58. **Niedergerke, R. and Page, S.,** Receptor controlled calcium discharge in frog heart cells, *Q. J. Exp. Physiol.,* 74, 987, 1989.
59. **Yasui, K., Palade, P., and Sandor, G.,** Negative control mechanism with features of adaptation controls Ca^{2+} release in cardiac myocytes, *Biophys. J.,* 67, 457, 1994.
60. **Franzini-Armstrong, C.,** Freeze-fracture of frog slow tonic fibers. Structure of surface and internal membranes, *Tissue Cell,* 16(3), 146, 1984.

Chapter 2

REGULATION OF THE CALCIUM RELEASE CHANNEL

Barbara E. Ehrlich and Andrew R. Marks

CONTENTS

1. INTRODUCTION

Changes in the intracellular calcium (Ca) concentration are used by virtually all cell types to trigger a vast array of cellular events including excitation-contraction coupling, oocyte fertilization, hormone secretion, neurotransmitter release, and T lymphocyte activation. Ca can enter the cytoplasm from the extracellular fluid through voltage- or ligand-gated channels, or from internal stores through intracellular release channels. Two major classes of intracellular Ca release channel have been identified, the ryanodine receptor (RyR)[1,2] and the inositol 1,4,5-trisphosphate receptor ($InsP_3R$).[3] Most cells contain both types of channel, but the relative densities vary dramatically. Cells also can have multiple isoforms of these channels.[4-6] The production of a variety of

0-8493-8543-1/95/$0.00+$.50

channels is not surprising as cells can respond to diverse stimuli with specific responses.

On the basis of cloning studies it is now appreciated that the Ca release channels of the sarcoplasmic and endoplasmic reticula comprise a distinct gene family which currently includes the two major channel types, the RyR and $InsP_3R$. There is approximately 40% homology between the RyR and $InsP_3R$ in some of the putative transmembrane regions—sufficient sequence similarity to suggest that these two channels evolved from a common ancestral cation release channel. The unifying characteristic of the Ca release channels is the fourfold symmetric structure of the channel, in which each channel is made of four identical subunits. The Ca release channels are also distinguished by their size; they are the largest ion channels identified to date. A third structural similarity is the clustering of the putative transmembrane segments near the carboxy terminus of each Ca release channel. These putative transmembrane segments are encoded by approximately 10% of the sequence, the remainder of the sequence encoding large cytoplasmic structures that presumably participate in regulating channel function and perhaps interact with other proteins.

This chapter will focus on the RyR. This receptor has been shown to be the same as the Ca release channel seen in vesicle[7] and bilayer experiments[8] and the same as the foot protein seen in electron micrographs.[9] The channel is a tetramer comprised of four 565,000 mol wt RyR subunits. The skeletal, cardiac, and brain forms share ~66% sequence homology. RyRs have been detected in nonmuscle tissues as well as in all types of muscle. Functional expression of the RyR has recently demonstrated that all of the important modulatory binding sites are encoded by the 16 kb RyR mRNA[10,11] and that a second protein, FKBP12, is required for optimal channel gating.[11]

Two models currently exist for the surface topography of the RyR. One model, from Numa's group, identifies 4 transmembrane segments located near the carboxy terminus,[12] the other, by MacLennan's group, suggests there are as many as 10 to 12 transmembrane segments.[13] The surface topography of the RyR provides the basis for designing strategies to dissect structure-function relationships. For example, one must know which regions are likely to be cytoplasmic and which are likely to be found within the SR in order to identify candidate modulator binding sites. Using protease sensitivity mapping,[14] six protease-sensitive regions that correspond to putative surface-exposed regions of the molecule were identified on the linear sequence of the ryanodine receptor (determined by cDNA cloning). Several regions were identified with high surface probability but protease resistance. These are candidates for sequences lining the channel pore(s), and for regions of protein-protein interaction during formation of the foot structure. Regions with high surface probability and protease sensitivity are candidates for modulator binding sites.

An important question is the identity of the physiological agonist(s) for the RyR. In skeletal muscle it is likely that there is mechanical coupling between the plasmalemmal Ca channel (the dihydropyridine receptor) and the RyR,[15] but the primary candidate for the agonist that opens the RyR in nonskeletal muscle is Ca. Ca may also contribute to activation of skeletal RyR that are not linked to dihydropyridine receptors.[16] Although one group suggested that cADP ribose could directly activate the RyR,[17] more recent reports provide compelling evidence that cADP ribose is unlikely to be an agonist for the RyR under physiological conditions.[18,19] In neurons it has been assumed that the RyR is activated by Ca, but there are conflicting reports. Evidence for Ca-induced Ca release exists for some cells such as cerebellar granule cells[20] but, for others, Ca release from intracellular stores has not been observed.

2. COMPARISON OF PUTATIVE ISOFORMS OF THE RYANODINE RECEPTOR/CHANNEL

The RyR has been found in virtually all cell types from a large variety of organisms including human heart,[21] scorpion skeletal muscle,[16] avian brain,[22] and sea urchin eggs.[23] It is remarkable that similarities in the primary sequence, the ultrastructural appearance, the pharmacology, and the electrophysiology make this protein recognizable as a single entity in all these cells. Nonetheless, there are interesting variations in individual properties among the channels. The channel complex appears to be modulated to suit the cellular function expected of the channel.

The "heavy" sarcoplasmic reticulum (SR) of skeletal muscle was the first preparation used to show Ca-gated channels from an intracellular organelle.[8] As shown by a number of laboratories, the skeletal receptor from a variety of mammalian species appears similar. Receptor from rabbit,[8,24] porcine,[25] and frog[26] skeletal muscle appears as a 120 pS channel (with 50 m*M* Ca as the charge carrier) with similar pharmacological characteristics. Despite differences in closing kinetics, channels from slow and fast twitch muscle have similar pharmacology.[27] Similarly, the effects of the chemotherapeutic agent, doxorubicin, are the same on RyR from canine and ovine muscle (compare References 28 and 29).

Despite the similarities in the single-channel conductance, evidence has been presented to suggest that there may be isoforms of this channel within a single tissue. Sutko and co-workers, for example, have reported that two foot proteins (labeled α and β) can be identified in avian skeletal muscle.[4,30] These two foot proteins differ in molecular weight, immunological cross-reactivity, and peptide maps.[4] Moreover, the expression of these two foot proteins changes during development.[30] A third foot protein isoform may also be present in these avian skeletal muscles.[30] These channel isoforms, coexisting in a single tissue, appear to be functionally distinct,[31,32] although the extent and importance of these structural differences needs further clarification.[33]

In contrast, the RyR from cardiac and skeletal muscle show differences at the level of the primary structure,[34] single-channel conductance,[24,35] and in sensitivity to Ca, ruthenium red, and other pharmacological agents (from a comparison of published reports and our unpublished observations at Ehrlich laboratory). The single-channel conductance of the cardiac channel (with 50 m*M* Ca as the current carrier) is 100 pS, compared to skeletal at 120 pS. From a comparison of the physiology of the two muscle types it was predicted that the channel from cardiac muscle would be more sensitive to Ca than the channel from skeletal muscle; this result has been demonstrated by some[35] but not all investigators. Additional functional divergencies will probably be found as comparisons of the primary sequence of the two channels are studied and manipulated.

The cardiac RyR from a variety of mammalian species has been studied, including canine,[24,35] porcine (unpublished observations at Ehrlich laboratory), ovine,[36] and human[21] heart. As with the skeletal muscle channel, the channel appears similar in all these species. Differences were found when we compared the Ca release channels from the intraventricular septum and the free wall of the canine left ventricle.[37] These results were not obtained when channel properties of the purified RyRs from the two regions of the heart were compared.[38] It is interesting to speculate that the functional differences observed in the native channels from the two regions of the heart may be a consequence of regional variation in the expression of FKBP isoforms. Skeletal muscle appears to contain one isoform of FKBP,[39] whereas cardiac muscle contains an additional isoform of this regulatory protein.[40]

As part of a study of Ca release channels from mammalian brain, we investigated the properties of the RyR from canine cerebellum. We initially estimated the slope conductance of this isoform of the channel to be 50 pS from current measurements at 0 mV;[41] more complete measurements show that the conductance is the same as the cardiac channel. The brain RyR is activated by caffeine, ATP, and Ca; it is inhibited by ruthenium red and modified by ryanodine.[41] We also found that the Ca dependence of the open probability is quite similar to that of the skeletal and cardiac RyR (compare References 41 to 43). It is interesting to note, however, that the brain RyR is more similar to the cardiac RyR than the skeletal receptor in primary sequence.[34]

A recent report identified a 2.4 kb transcript of the skeletal muscle RyR that was expressed in rabbit brain.[12] These authors have proposed that this short transcript forms a Ca release channel comprised of only the carboxy terminal transmembrane regions encoded by the much larger 16 kb RyR transcript. Such a channel might lack certain modulatory binding sites found in the much larger Ca release channels of striated muscles. Further analysis of this truncated version of the RyR may provide important insights into the structure/function relationship of this channel.

In bilayer experiments we also observed the RyR from the ER of two other tissues, scallop muscle and sea urchin egg. In both cases the channel conductance with 50 m*M* Ca as the charge carrier was 80 pS, the channel was activated by caffeine, ATP, and Ca, and was inhibited by ruthenium red. These observations with the channels from scallop and sea urchin are interesting because, although the invertebrates and vertebrates diverged about 50 million years ago,[44] the Ca release channels appear remarkably similar. Certainly the structure of the channel complex appears similar when analyzed by electron microscopy.[16] In contrast, the RyR from lobster skeletal muscle appears to be relatively insensitive to ATP, Ca, and Mg when compared to the mammalian RyR, despite the ability to measure specific ryanodine binding to a high molecular weight protein that has many other properties similar to the mammalian RyR.[45]

3. FUNCTIONAL EXPRESSION OF THE SKELETAL MUSCLE RYANODINE RECEPTOR/CHANNEL

The primary sequences of the RyR from skeletal muscle,[12,13,46] cardiac muscle,[34,47] and brain[48] have been deduced from cDNA cloning. The cloned RyR from skeletal and cardiac muscle has been expressed in CHO cells and in *Xenopus* oocytes by Numa and colleagues.[34,49] These studies have demonstrated that each of these RyR cDNAs encode the subunit of the Ca release channel in cardiac and skeletal muscle. More recently, single-channel recordings of the cloned skeletal RyR expressed in COS-1 cells have been reported, however, multiple conductances were observed from these cloned expressed channels including some channels with conductances >1 nS, which are not seen with native ryanodine receptors.[10]

We have expressed the skeletal muscle RyR in insect cells and have shown that these recombinant channels behave exactly like the native ryanodine receptor in terms of activation and inhibition by channel modulators, conductance, and cation selectivity (Figure 1).[11] First, to show that the RyR cDNA encoded a functional Ca release channel, *in vitro* transcribed RyR mRNA was injected into *Xenopus* oocytes.[11] RyR expression was determined by recording the native Ca-activated Cl^- current of *Xenopus* oocytes using voltage clamp techniques. Addition of caffeine induced an inward current, and after washout of the caffeine, the current decayed to baseline levels. The current amplitude was unchanged in the absence of extracellular Ca, indicating that the Ca used to activate these currents came from intracellular stores. Cells that were not injected with RyR mRNA did not respond to caffeine. This series of experiments verified that a caffeine-sensitive intracellular Ca release channel was expressed in oocytes injected with RyR mRNA.

To obtain cloned RyR suitable for reconstitution in lipid bilayers, Sf9 cells were infected with recombinant Baculovirus containing the RyR cDNA. Northern analysis of total RNA isolated from infected Sf9 cells revealed a 16 kb mRNA which hybridized to an α-^{32}P- labeled RyR cDNA. This demonstrated that the infected Sf9 cells were expressing full-length RyR mRNA. We chose to study purified cloned expressed RyR to eliminate the effects of contaminating proteins that could contribute to modifying Ca release channel function in poorly defined ways. Cloned expressed RyR was purified from Sf9 cells on a

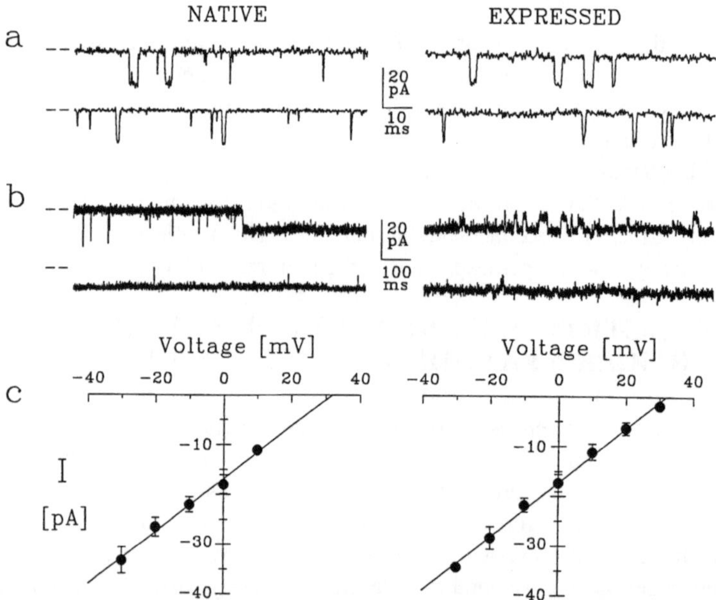

FIGURE 1. Comparison of single-channel recording from purified native and recombinant RyRs expressed in insect cells. Native and cloned expressed RyRs were isolated under identical conditions using sucrose density purification, incorporated into liposomes, and fused to planar lipid bilayers. (**a**) Channels with high conductances for Cs were observed in both native and cloned expressed preparations. (**b**) Both native and expressed RyRs were modulated by ryanodine (0.5 μ*M*). Note in the top trace of (**b**) for the native RyR that midway through this trace the channel opens to the reduced conductance state characteristically induced by ryanodine. The amplitude of the single channel currents at 0 mV was 18.3 ± 1.3 pA before the addition of ryanodine and 9 ± 0.4 pA after the addition of ryanodine. Channel opening and closing kinetics were slowed after the addition of ryanodine (note that the time scale is decreased tenfold in the presence of ryanodine). (**c**) The current-voltage relationship of the Cs current through the native and purified recombinant RyR reconstituted into bilayers. The slope conductances were 508 ± 35 pS for the native channel and 540 ± 48 for the expressed channel. In all cases dashed lines on the left of each tracing represent the closed state, channel openings are downward.

sucrose gradient using ^{3}H-ryanodine to monitor the degree of purification. After reconstitution into planar lipid bilayers, recombinant RyR formed a channel with properties indistinguishable from those of the native RyR. This channel exhibited a conductance of 540 ± 48 pS (with Cs 250:50 m*M* as the charge carrier) and was modulated to the characteristic subconductance state by ryanodine.[11] Recombinant RyRs expressed in Sf9 cells formed a channel that was Ca sensitive (activated by 100 μ*M* Ca) and was activated by 1 m*M* ATP, and inhibited by 20 μ*M* ruthenium red and by 1 m*M* $MgCl_2$.[11] Single-channel currents could also be observed with Ca as the current carrier; in this case the conductance was 110 pS, as observed in experiments using native RyRs.

4. FKBP12 OPTIMIZES THE FUNCTION OF THE RYANODINE RECEPTOR/CHANNEL

The Ca release channel/RyR has previously been reported to consist of four identical 565 kDa protomers. We recently reported the identification of a 12 kDa protein that is tightly associated with highly purified RyR from rabbit skeletal muscle SR.[39] N-terminal amino acid sequencing followed by cDNA cloning and sequencing demonstrated that the 12 kDa protein from skeletal muscle is the binding protein for the immunosuppressant drug FK506. The FK506 binding protein (FKBP12) is an immunophilin originally identified by its ability to bind to the structurally related immunosuppressant drugs FK506 and rapamycin.[50]

This accessory protein, FKBP12, co-purifies with the RyR during sucrose density gradient centrifugation, co-localizes to the terminal cisternae of the SR, and an anti-FKBP12 antibody can immunoprecipitate the RyR.[39] FKBP12, a *cis-trans* peptidyl-prolyl isomerase, binds the immunosuppressant drugs FK506 and rapamycin.[50] However, the prolyl isomerase activity of FKBP12 is not required for the immunosuppressant properties of the two drugs.[51] The rapamycin-FKBP12 complex blocks the cell cycle transition from G1 to S[39,52-57] and inhibits p70 S6 and p34cdc2 kinases.[39,57,58] The target for the rapamycin-FKBP12 complex has been identified as a mammalian homologue of the yeast TOR proteins.[59,60]

The physiological function of FKBP12 in the absence of the immunosuppressant drugs was unclear; the association between the RyR and FKBP12[39] and the potency of FKBP12 as a chemoattractant[61] have suggested important cellular functions. In addition, the findings that there is a family of FKBPs and that these binding proteins are found in virtually all cell types[54] suggest that these compounds are involved in a range of cellular processes.

We found that FKBP12 and the RyR are tightly associated in skeletal muscle SR on the basis of (1) co-purification through sequential heparin agarose, hydroxylapatite, and size exclusion chromatography columns; (2) co-immunoprecipitation of the RyR and FKBP12 with anti-FKBP12 antibodies; and (3) subcellular localization of both proteins to the terminal cisternae of the SR, not the longitudinal tubules of SR, in skeletal muscle.[39] The molar ratio of FKBP12 to RyR in highly purified RyR preparations is approximately one to one,[62] indicating that one FKBP12 molecule is associated with each subunit of the Ca release channel/RyR (Figure 2).

We have recently shown that FKBP12 modulates Ca release channel gating.[11] By co-expressing the RyR FKBP12 in insect cells, we demonstrated that FKBP12 optimizes channel behavior and that this effect can be reversed by adding FK506 or rapamycin, both of which inhibit the isomerase activity of FKBP12.[11] These results provided the first cellular function for FKBP12, and established that the functional Ca release channel is a complex comprised of

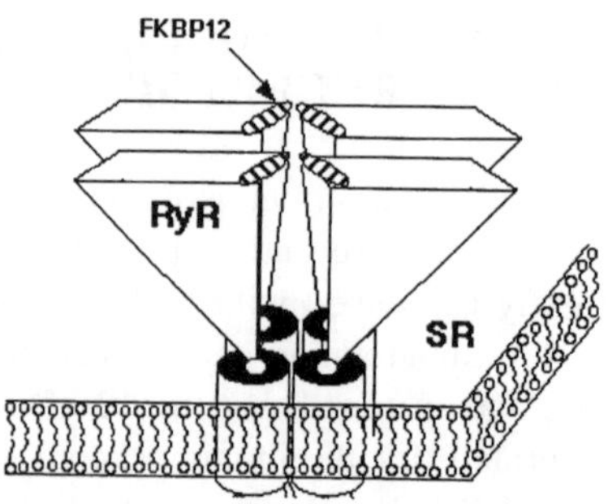

FIGURE 2. Model for the RyR channel complex.

RyR and FKBP12. The fact that the drugs rapamycin and FK506, both of which inhibit the isomerase activity of FKBP12, reverse the channel optimizing effects of FKBP12 suggests that conserved prolines may be important in regulating channel function.

Similarly, we have examined the effects of rapamycin on the function of the native RyR.[63] The related compound, FK506, can dissociate FKBP12 from RyR[62] and alters the function of the skeletal muscle by increasing open probability.[64,65] We found that rapamycin altered channel activity by increasing the open probability (Po) of the channel from both types of muscle; in addition, rapamycin decreased the single-channel conductance of the Ca release channel from cardiac muscle.[63] Rapamycin added to the cloned, expressed skeletal RyR, which does not include FKBP12,[11] had no effect on channel function. This finding indicates that the effects of these immunosuppressant drugs are dependent upon the presence of FKBP12, not direct effects on the RyR itself. These experiments demonstrate that FKBP12 plays a functional role in the activity of both cardiac and skeletal muscle SR Ca release channels.

5. REGULATION OF RYANODINE RECEPTOR EXPRESSION IN HUMAN HEART FAILURE

Abnormal Ca homeostasis has been demonstrated in failing human myocardium.[66] Previous studies found that expression of the SR Ca-ATPase and dihydropyridine receptor were decreased in end-stage human heart failure.[67,68] Recent work has demonstrated that expression of the RyR was also reduced in the myopathic left ventricle compared to normal.[69,70]

Previous studies that examined Ca channel expression in human myocardium focused on the left ventricle. Virtually no information was available on the expression and regulation of the Ca channels in other regions of the human heart. Right ventricular endomyocardial biopsies are generally used as a source of human myocardium for the characterization of cardiac muscle-specific gene expression with the assumption that gene expression is uniform in all regions of the heart.

We found that RyR mRNA expression in the left ventricle and the interventricular septum were decreased by 30% in patients with heart failure.[71] Interestingly, there was an opposite trend of increased RyR mRNA expression in the myopathic right ventricle compared with normals, albeit the increase was statistically nonsignificant.[71] The absence of a significant difference in RyR expression between normal and myopathic right ventricle may be partly due to the varying degree of right ventricular failure present among the patients. In the same samples the levels of $InsP_3R$ expression were increased significantly in myopathic heart tissue in the three regions of the heart tested (left and right ventricles and interventricular septum) when compared with normal tissue.[71] Radioligand binding studies showed that there was less ryanodine binding, but more $InsP_3$ binding, in cardiomyopathic tissue compared with normal.[71]

The decreased expression in cardiomyopathic left ventricles of the two channels involved in excitation-contraction coupling, the dihydropyridine receptor and the RyR, implies that intracellular Ca regulation via this pathway may be impaired and could contribute to compromised cardiac contractility. In contrast, there is a consistent trend of increased $InsP_3R$ mRNA expression in failing human myocardium compared to normal, specifically in both ventricles. An alteration in the expression of Ca handling molecules may be a molecular mechanism underlying impaired excitation-contraction coupling, compensatory hypertrophy, enhanced pharmacomechanical coupling, and increased arrhythmia susceptibility, all of which are observed in progressive myocardial dysfunction. These insights into the molecular basis of heart failure should provide new directions for future diagnostic and therapeutic approaches.

6. CONCLUSIONS

The RyR is a ligand-gated Ca-permeable channel associated with intracellular membranes. Activation of this channel protein opens a pore for Ca, which can then diffuse from the lumen of the sarcoplasmic reticulum to the cytoplasm. The RyR is regulated by a number of cellular pathways and by numerous pharmacological agents.[72] In this review several physiologically important functional properties of the RyR have been described.

Comparisons of the function of the RyR from a variety of tissues and organisms provides initial insights into the relationship between the structure and function of this intracellular channel. As these comparisons are combined with observations garnered from the other member of the gene family, the $InsP_3R$, a cohesive picture of the evolution of the cation release channels will emerge.

The functional expresssion of the cloned RyR revealed that the protein was necessary and sufficient to generate Ca release channels. It was notable to observe that virtually all of the regulatory sites are present on the RyR protein. Nonetheless, an associated protein, FKBP12, has profound effects on the gating of the channel, suggesting that the native channel complex is comprised of at least these two protein types.

Information about the structure, heterogeneity, functional properties, and regulation of the RyR will allow an in-depth understanding of Ca signaling by cells. The combination of biochemical, biophysical, and molecular biological approaches have begun to reveal the secrets of the RyR. Future investigations will further our understanding of the role and regulation of this receptor/channel in health and disease.

ACKNOWLEDGMENTS

We thank Drs. Loewe Go, Edward Kaftan, and Karol Ondrias for their expertise in the experiments presented and for comments on the manuscript. This work was supported by NIH grants HL 33026 (BEE), NS 29814 (ARM) and Grant-in-Aids from the Patrick and Catherine Weldon Donaghue Medical Research Foundation (BEE) and the American Heart Association (ARM).

REFERENCES

1. **Lai, F. A. and Meissner, G.,** The muscle ryanodine receptor and its intrinsic Ca channel activity, *J. Bioenerg. Biomembr.*, 55, 227, 1989.
2. **Fleischer, S. and Inui, M.,** Biochemistry and biophysics of excitation-contraction coupling, *Annu. Rev. Biophys. Biophys. Chem.*, 18, 333, 1989.
3. **Berridge, M. J.,** Inositol trisphosphate and calcium signaling, *Nature*, 361, 315, 1993.
4. **Airey, J. A., Beck, C. F., Murakami, K., Tanksley, S. J., Deerinck, T. J., Ellisman, M. H., and Sutko, J. L.,** Identification and localization of 2 triad junctional foot protein isoforms in mature avian fast twitch skeletal muscle, *J. Biol. Chem.*, 265, 14187, 1990.
5. **Sudhof, T. C., Newton, C. L., Archer, B. T., Ushkaryov, Y. A., and Mignery, G. A.,** Structure of a novel $InsP_3$ receptor, *EMBO J.*, 10, 3199, 1991.
6. **Danoff, S. K., Ferris, C. D., Donath, C., Fischer, G. A., Munemitsu, S., Ullrich, A., Snyder, S. H., and Ross, C. A.,** Inositol 1,4,5-trisphosphate receptors: distinct neuronal and nonneuronal forms derived by alternative splicing differ in phosphorylation, *Proc. Natl. Acad. Sci. U.S.A.*, 88, 2951, 1991.
7. **Smith, J. S., Imagawa, T., Ma, J., Fill, M., Campbell, K. P., and Coronado, R.,** Rurified ryanodine receptor from rabbit skeletal muscle is the calcium release-channel of sarcoplasmic reticulum, *J. Gen. Physiol.*, 92, 1, 1988.
8. **Smith, J., Coronado, R., and Meissner, G.,** Single channel measurements of the calcium release channel from skeletal muscle sarcoplasmic reticulum. Activation by Ca^{2+} and ATP and modulation by Mg^{2+}, *J. Gen. Physiol.*, 88, 573, 1986.
9. **Franzini-Armstrong, C.,** Studies of the triad. I. Structure of the junction in frog twitch fibers, *J. Cell Biol.*, 47, 488, 1970.
10. **Chen, S. R. W., Vaughan, D. M., Airey, J. A., Coronado, R., and MacLennan, D. H.,** Functional expression of cDNA encoding the Ca release channel (ryanodine receptor) of rabbit skeletal muscle sarcoplasmic reticulum in COS-1 cells, *Biochemistry*, 32, 3743, 1993.
11. **Brillantes, A.-M. B., Ondrias, K., Scott, A., Kobrinsky, E., Ondriasova, E., Moschella, M. C., Jayaraman, T., Landers, M., Ehrlich, B. E., and Marks, A. R.,** Stabilization of calcium release channel (ryanodine receptor) function by FK-506 binding protein, *Cell*, 77, 513, 1994.

12. **Takeshima, H., Nishimura, S., Matsumoto, T., Ishida, H., Kangawa, K., Minamino, N., Matsuo, H., Ueda, M., Hanaoka, M., Hirose, T., and Numa, S.,** Primary structure and expression from complementary DNA of skeletal muscle ryanodine receptor, *Nature*, 339, 439, 1989.
13. **Zorzato, F., Fujii, J., Otsu, K., Phillips, M., Green, N. M., Lai, F. A., Meissner, G., and MacLennan, D. H.,** Molecular cloning of cDNA encoding human and rabbit forms of the Ca release channel (ryanodine receptor) of skeletal muscle sarcoplasmic reticulum, *J. Biol. Chem.*, 265, 2244, 1990.
14. **Marks, A. R., Fleischer, S., and Tempst, P.,** Surface topography analysis of the ryanodine receptor/junctional channel complex based on proteolysis sensitivity mapping, *J. Biol. Chem.*, 265, 13143, 1990.
15. **Rios, E., Ma, J., and Gonzalez, A.,** The mechanical hypothesis of excitation-contraction (EC) coupling in skeletal muscle, *J. Muscle Res. Cell Motil.*, 12, 127, 1991.
16. **Loesser, K. E., Castellani, L., and Franzini-Armstrong, C.,** Dispositions of junctional feet in muscles of invertebrates, *J. Muscle Res. Cell Motil.*, 13, 161, 1992.
17. **Meszaros, L. G., Bak, J., and Chu, A.,** Cyclic ADP-ribose as an endogenous regulator of the non-skeletal type ryanodine receptor Ca^{2+} channel, *Nature*, 364, 76, 1993.
18. **Sitsapesan, R., McGarry, S. J., and Williams, A. J.,** Cyclic ADP-ribose competes with ATP for the adenine nucleotide binding site on the cardiac ryanodine receptor Ca^{2+}-release channel, *Circ. Res.*, 75, 596, 1994.
19. **Morrissette, J., Heisermann, G., Cleary, J., Ruoho, A., and Coronado, R.,** Cyclic ADP-ribose induced Ca release in rabbit skeletal muscle sarcoplasmic reticulum, *FEBS Lett.*, 330, 270, 1993.
20. **Whitham, E. M., Challiss, R. A. J., and Nahorski, S. R.,** Inositol 1,4,5-trisphosphate-stimulated calcium release from permeabilized cerebellar granule cells, *Br. J. Pharmacol.*, 104, 202, 1991.
21. **Holmberg, S. R. M. and Williams, A. J.,** Single channel recordings from human cardiac sarcoplasmic reticulum, *Circ. Res.*, 65, 1445, 1989.
22. **Ellisman, M. H., Deerinck, T. J., Ouyang, Y., Beck, C. F., Tanksley, S. J., Walton, P. D., Airey, J. A., and Sutko, J. L.,** Identification and localization of ryanodine binding proteins in the avian central nervous system, *Neuron*, 5, 135, 1990.
23. **McPherson, S. M., McPherson, P. S., Campbell, K. P., and Longo, F. J.,** Localization of a calcium release channel in sea urchin eggs, *J. Cell Biol.*, 115, 321, 1991.
24. **Ehrlich, B. E. and Watras, J.,** Inositol 1,4,5-trisphosphate activates a channel from smooth muscle sarcoplasmic reticulum, *Nature*, 336, 583, 1988.
25. **Ervasti, J. M., Strand, M. A., Hanson, T. P., Mickelson, J. R., and Louis, C. F.,** Ryanodine receptor in different malignant hyperthermia-susceptible porcine muscles, *Am. J. Physiol.*, 260, C58, 1991.
26. **Suarez-Isla, B. A., Alcayaga, C., Marengo, J. J., and Bull, R.,** Activation of inositol trisphosphate-sensitive Ca^{2+} channels of sarcoplasmic reticulum from frog skeletal muscle, *J. Physiol. London*, 441, 575, 1991.
27. **Lee, Y. S., Ondrias, K., Duhl, A. J., Ehrlich, B. E., and Kim, D. H.,** Comparison of calcium release from sarcoplasmic reticulum of slow and fast twitch muscles, *J. Membr. Biol.*, 122, 155, 1991.
28. **Ondrias, K., Borgatta, L., Kim, D. H., and Ehrlich, B. E.,** Biphasic effects of doxorubicin on the calcium release channel from sarcoplasmic reticulum of cardiac muscle, *Circ. Res.*, 67, 1167, 1990.
29. **Holmberg, R. M. and Williams, A. J.,** Patterns of interaction between anthraquinone drugs and the calcium-release channel from cardiac sarcoplasmic reticulum, *Circ. Res.*, 67, 272, 1990.
30. **Sutko, J. L., Airey, J. A., Murakami, K., Takeda, M., Beck, C., Deerinck, T., and Ellisman, M. H.,** Foot protein isoforms are expressed at different times during embryonic chick skeletal muscle development, *J. Cell Biol.*, 113, 793, 1991.

31. **Percival, A. L., Williams, A. J., Kenyon, J. L., Grinsell, M. M., Airey, J. A., and Sutko, J. L.,** Chicken skeletal muscle ryanodine receptor isoforms: ion channel properties, *Biophys. J.*, 67, 1834, 1994.
32. **Valdivia, H. H., Fuentes, O., and Block, B.,** *Biophys. J.*, 66, A418, 1994.
33. **Oyamada, H. et al.,** Primary structure and distribution of ryanodine-binding protein isoforms of the bullfrog skeletal muscle, *J. Biol. Chem.*, 269, 17206, 1994.
34. **Nakai, J., Imagawa, T., Hakamata, Y., Shigekawa, M., Takeshima, H., and Numa, H.,** Primary structure and functional expression from cDNA of the cardiac ryanodine receptor/calcium release channel, *FEBS Lett.*, 271, 169, 1990.
35. **Rousseau, E., Smith, J. S., Henderson, J. S., and Meissner, G.,** Single channel and $^{45}Ca^{2+}$ flux measurements of the cardiac sarcoplasmic reticulum calcium channel, *Biophys. J.*, 50, 1009, 1986.
36. **Lindsay, A. R. G., Manning, S. D., and Williams, A. J.,** Monovalent cation conductance in the ryanodine receptor-channel of sheep cardiac muscle sarcoplasmic reticulum, *J. Physiol.*, 439, 463, 1991.
37. **Borgatta, L., Watras, J., Katz, A. M., and Ehrlich, B. E.,** Regional differences in calcium-release channels from heart, *Proc. Natl. Acad. Sci. U.S.A.*, 88, 2486, 1991.
38. **Xu, L., Cohn, A., and Meissner, G.,** Ryanodine sensitive calcium release channel from left ventricle, septum, and atrium of canine heart, *Cardiovasc. Res.*, 27, 1815, 1993.
39. **Jayaraman, T., Brillantes, A. M., Timerman, A. P., Fleischer, S., Erdjumentbromage, H., Tempst, P., and Marks, A. R.,** FK506 binding protein associated with the calcium release channel (ryanodine receptor), *J. Biol. Chem.*, 267, 9474, 1992.
40. **Timerman, A. P., Jayaraman, T., Wiederrecht, G., Onoue, H., Marks, A. R., and Fleischer, S.,** The ryanodine receptor from canine heart sarcoplasmic reticulum is associated with a novel FK-506 binding protein, *Biochem. Biophys. Res. Commun.*, 198, 701, 1994.
41. **Bezprozvanny, I., Watras, J., and Ehrlich, B. E.,** Bell-shaped calcium-response curves of Ins(1,4,5)P_3- and calcium-gated channels from endoplasmic reticulum of cerebellum, *Nature*, 351, 751, 1991.
42. **Meissner, G., Darling, E., and Eveleth, J.,** Kinetics of rapid Ca^{2+} release by sarcoplasmic reticulum. Effects of Ca^{2+}, Mg^{2+} and adenine nucleotides, *Biochemistry*, 25, 236, 1986.
43. **Kawano, S. and Coronado, R.,** Ca^{2+} dependence of Ca^{2+} release channel activity in the sarcoplasmic reticulum of cardiac and skeletal muscle, *Biophys. J.*, 59, 600, 1991.
44. **Hille, B.,** *Ionic Channels of Excitable Membranes*, Sinauer Associates, Sunderland, MA, 1992.
45. **Seok, J.-H., Xu, L., Kramarcy, N. R., Sealock, R., and Meissner, G.,** The 30 S lobster skeletal muscle Ca release channel (ryanodine receptor) has functional properties distinct from the mammalian channel proteins, *J. Biol. Chem.*, 267, 15893, 1992.
46. **Marks, A. R., Tempst, P., Hwang, K. S., Taubman, M. B., Inui, M., Chadwick, C., Fleischer, S., and Nadal-Ginard, B.,** Molecular cloning and characterization of the ryanodine receptor/junctional channel complex cDNA from skeletal muscle sarcoplasmic reticulum, *Proc. Natl. Acad. Sci. U.S.A.*, 86, 8683, 1989.
47. **Otsu, K., Willard, H. F., Khanna, V. K., Zorzato, F., Green, N. M., and MacLennan, D. H.,** Molecular cloning of cDNA encoding the Ca^{2+} release channel (ryanodine receptor) of rabbit cardiac muscle sarcoplasmic reticulum, *J. Biol. Chem.*, 265, 13472, 1990.
48. **Hakamata, Y., Nakai, J., Takaashima, H., and Imoto, K.,** Primary structure and distribution of a novel ryanodine receptor/Ca release channel from rabbit brain, *FEBS Lett.*, 312, 229, 1992.
49. **Penner, R., Neher, E., Takeshima, H., Nishimura, S., and Numa, S.,** Functional expression of the calcium release channel from skeletal muscle ryanodine receptor cDNA, *FEBS Lett.*, 259, 217, 1989.

50. **Harding, M. W., Galat, A., Uehling, D. E., and Schreiber, S. L.,** A receptor for the immunosuppressant FK506 is a cis-trans peptidyl-prolyl isomerase, *Nature*, 341, 758, 1989.
51. **Bierer, B., Somers, P. K., Wandless, T. J., Burakoff, S. J., and Schreiber, S. L.,** Probing immunosuppressant action with a nonnatural immunophilin ligand, *Science*, 250, 556, 1990.
52. **Bierer, B. E., Mattilia, P. S., Standaert, R., Herzenberg, L., Burakoff, S., Crabtree, G., and Schreiber, S.,** Two distinct signal transmissionn pathways in T lymphocytes are inhibited by complexes formed between an immunophilin and either FK506 or rapamycin, *Proc. Natl. Acad. Sci. U.S.A.*, 87, 9231, 1990.
53. **Dumont, F. J., Staruch, M. J., Kooprak, S. L., Melino, M. R., and Sigal, N. H.,** Distinct mechanisms of suppression of murine T-cell activation by the related macrolides FK-506 and rapamycin, *J. Immunol.*, 144, 251, 1990.
54. **Schreiber, S.,** Chemistry and biology of the immunophilins and their immunosuppressive ligands, *Science*, 251, 283, 1991.
55. **Chung, J., Kuo, C. J., Crabtree, G. R., and Blenis, J.,** Rapamycin-FKBP specifically blocks growth-dependent activation of and signaling the 70 kDa S6 protein kinases, *Cell*, 69, 1227, 1992.
56. **Kuo, C. J., Chung, J., Fiorentino, D. F., Flanagan, W. M., Blenis, J., and Crabtree, G. R.,** Rapamycin selectivelly inhibits interleukin-2 activation and p70 S6 kinase 1, *Nature*, 358, 70, 1992.
57. **Price, D. J., Grove, J. R., Calvl, V., Avruch, J., and Bierer, B. E.,** Rapamycin-induced inhibition of the 70-kilodalton S6 protein kinase, *Science*, 257, 1992.
58. **Morice, W. G., Brunn, G. J., Wiederrecht, G., Siekierka, J. J., and Abraham, R. T.,** Rapamycin-induced inhibition of p34cdc2 kinase activation is associated with G1/S-phase growth arrest in T lymphocytes, *J. Biol. Chem.*, 268, 3734, 1993.
59. **Brown, E., Albers, M., Shin, T., Ichikawa, K., Keith, C., Lane, W., and Schreiber, S.,** A mammalian protein targeted by G1-arresting rapamycin-receptor conplex, *Nature*, 369, 756, 1994.
60. **Sabatini, D., Erdjument-Bromage, H., Lui, M., Tempst, P., and Snyder, S.,** RAFT1: a mammalian protein that binds FKBP12 in a rapamycin-dependent fashion and is homologous to yeast TORs, *Cell*, 78, 35, 1994.
61. **Leiva, M. C. and Lyttle, C. R.,** Leukocyte chemotactic activity of FKBP and inhibition by FK506, *Biochem. Biophys. Res. Commun.*, 186, 1178, 1992.
62. **Timerman, A. P., Ogunbunmi, E., Freund, E., Wiederrecht, G., Marks, A., and Fleischer, S.,** The calcium release channel of sarcoplasmic reticulum is modulated by FK-506 binding protein, *J. Biol. Chem.*, 268, 22992, 1993.
63. **Kaftan, E., Marks, A. R., and Ehrlich, B. E.,** Effects of rapamycin on ryanodine receptor/calcium release channels from skeletal and cardiac muscle, submitted, 1995.
64. **Mayrleitner, M., Timerman, A. P., Wiederrecht, G., and Fleischer, S.,** The calcium release channel of sarcoplasmic reticulum is modulated by FK-506 binding protein: effect of FKBP-12 on single channel activity of the skeletal muscle ryanodine receptor, *Cell Ca.*, 15, 99, 1994.
65. **Ahern, G. P., Junankar, P. R., and Dulhunty, A. F.,** Single channel activity of the ryanodine receptor calcium release channel is modulated by FK-506, *FEBS Lett.*, 352, 369, 1994.
66. **Gwathmey, J. K., Copelas, L., McKinnon, R., Schoen, F., Feldman, M., Grossman, W., and Morgan, J. P.,** Abnormal intracellular calcium handling in myocardium from patients with end-stage heart failure, *Circ. Res.*, 61, 70, 1987.
67. **Mercadier, J.-J., Lompre, A.-M., Duc, P., Boheler, K. R., Fraysse, J.-B., Wisnewsky, C., Allen, P. D., Komajda, M., and Schwartz, K.,** Altered sarcoplasmic reticulum Ca-ATPase gene expression in the human ventricle during end-stage heart failure, *J. Clin. Invest.*, 85, 305, 1990.

68. **Takahashi, T., Allen, P. D., Marks, A. R., Denniss, A. R., Schoen, F. J., Grossman, W., Marsh, J. D., and Izumo, S.,** Altered expression of genes encoding the Ca^{2+} regulatory proteins in the myocardium of patients with end-stage heart failure: correlation with expression of the Ca^{2+}-ATPase gene, *Circ. Res.*, 71, 1357, 1992.
69. **Arai, M., Alpert, N. R., MacLennan, D. H., Barton, P., and Periasamy, M.,** Alterations in sarcoplasmic reticulum gene expression in human heart failure: a possible mechanism for alterations in systolic and diastolic properties of the failing myocardium, *Circ. Res.*, 72, 463, 1993.
70. **Brillantes, A.-M., Allen, P., Takahashi, T., Izumo, S., and Marks, A. R.,** Differences in cardiac calcium release channel (ryanodine receptor) expression in myocardium from patients with end-stage heart failure caused by ischemic versus dilated cardiomyopathy, *Circ. Res.,* 71, 18, 1992.
71. **Go, L. O., Moschella, M. C., Watras, J., Handa, K. K., Fyfe, B. S., and Marks, A. R.,** Differential regulation of two types of intracellular calcium release channels during end-stage human heart failure, *J. Clin. Invest.,* 95, 888, 1995.
72. **Coronado, R., Morrissette, J., Sukhareva, M., and Vaughan, D. M.,** Structure and function of ryanodine receptors, *Am. J. Physiol.,* 266, C1485, 1994.

Chapter 3

CYCLIC ADP-RIBOSE: A MEDIATOR OF A CALCIUM SIGNALING PATHWAY

Hon Cheung Lee

CONTENTS

1. INTRODUCTION

Multiple pathways are present in cells to transduce chemical information of the external environment to intracellular responses. The two best characterized are mediated by cyclic nucleotides and Ca^{2+}. The identification more than 35 years ago of cAMP as the mediator of the effects of epinephrine on glycogen breakdown has established the concept of chemical second messengers.[1] It is known now that the main action of cAMP is stimulation of cAMP-dependent kinase which, through protein phosphorylation, produces long-lasting modulations of cellular activities. The function of cGMP, another member of the cyclic nucleotide family, is more diverse. In addition to protein phosphorylation through cGMP-dependent kinase, cGMP has also been shown to affect ion channel activity directly[2] and regulate cyclic nucleotide phosphodiesterase.[3] Neither cAMP nor cGMP appears to be directly involved in Ca^{2+}-signaling.

Similar to the cyclic nucleotide pathway, binding of external ligands to surface receptors can also trigger elevation of internal Ca^{2+} due to influx and/or mobilization of internal Ca^{2+} stores. The two major mechanisms of Ca^{2+} mobilization, which are the main focus of this article, are the pathways mediated by inositol 1,4,5-trisphosphate (IP_3) and the Ca^{2+}-induced Ca^{2+} release mechanism (CICR). IP_3 is produced by receptor-activated hydrolysis of

0-8493-8543-1/95/$0.00+$.50

phosphoinositide bisphosphate, a minor phospholipid in the plasma membrane. IP_3 binds to a specific receptor in the endoplasmic reticulum and triggers Ca^{2+} release from the store (reviewed in Reference 4). The CICR mechanism is mediated by the ryanodine receptor (RyR), which is a Ca^{2+}-activated Ca^{2+} channel sharing considerable sequence homology with the IP_3-receptor.[5] Although exogenous substances such as ryanodine and caffeine can modulate the channel activity of the RyR, it is not known whether endogenous regulators of the receptor exist.

Cyclic ADP-ribose is a newly identified member of the cyclic nucleotide family.[6-8] Unlike the other members of the family, it is derived from NAD^+ instead of nucleotide triphosphates. It is naturally occurring in various mammalian tissues[9] and its signaling function appears to be mainly through mobilizing internal Ca^{2+} stores. As will be described in this article, increasing evidence indicates it may be an endogenous regulator of CICR. With the addition of this novel member, the signaling functions of the cyclic nucleotide family are now extended from protein phosphorylation to mobilization of internal Ca^{2+}.

The multiplicity of signaling pathways in cells generally do not operate in isolation but, instead, are interwoven to form elaborate information networks. This is the case for the IP_3 and the cAMP pathways. Through protein phosphorylation, the cAMP pathway can enhance the Ca^{2+} channel activity of the IP_3 receptor and thereby modulate the Ca^{2+} signaling pathway.[10] Similar cross-talk is observed between the cADPR and cGMP pathways, as the enzymatic synthesis of cADPR is shown to be stimulated by cGMP.[11] The two Ca^{2+} signaling pathways, therefore, are regulated by two different cyclic nucleotides, a noteworthy symmetry.

That cGMP is involved in the cADPR pathway raises the possibility that it is related to yet another novel signaling pathway that is mediated by nitric oxide (NO). NO was discovered as a smooth muscle relaxant produced by vascular endothelial cells (reviewed in References 12 and 13). It is a gaseous molecule freely permeable to the cell membrane and its signaling function is rapidly terminated by oxidation under physiological conditions. These properties make it ideally suited to be a messenger mediating intercellular information transfer. The main target of NO is the soluble guanylate cyclase in cells. By binding to the heme group, NO activates the enzyme resulting in elevation of intracellular cGMP. NO has been proposed to be a messenger for mediating a variety of other cell functions as well as vascular dilation (reviewed in References 12 and 13).

In addition to regulation by cGMP and NO, the cADPR pathway also has built-in feedback controls, being inhibited by desensitization and stimulated by Ca^{2+} and calmodulin. The wealth of controlling points of the pathway indicates it is tightly regulated. Although the understanding of the pathway is far from complete, the number of novel features that have been elucidated since its discovery in 1987 indicates it is emerging as one of the most exciting areas of investigation.

2. THE CYCLIC ADP-RIBOSE-DEPENDENT CALCIUM SIGNALING PATHWAY

Cyclic ADP-ribose was discovered as a Ca^{2+}-mobilizing metabolite produced during incubation of sea urchin egg lysates with NAD^+.[6] Egg homogenates, when prepared appropriately, contain functional microsomes and a variety of other organelles, providing a good representation of the Ca^{2+} regulation systems in intact eggs.[14] The microsomes possess an ATP-dependent Ca^{2+} transport system which can sequester contaminating Ca^{2+} in the medium and effectively lower the ambient Ca^{2+} concentration to the sensitive levels of Ca^{2+} indicator dyes. The self-loading feature and the remarkable stability of the egg homogenates make them a convenient and reliable bioassay for Ca^{2+} release activators.[14] Fractionation of the homogenates by Percoll density centrifugation shows that the microsomes are components of the endoplasmic reticulum, since the Ca^{2+} sequestering activity co-purifies with the marker glucose-6-phosphatase.[6,14,15]

Addition of IP_3 to egg microsomes elicits immediate and rapid release of Ca^{2+}. NAD^+ also induces Ca^{2+} release but with a different kinetic, showing a prominent initial delay.[6] Preincubation of NAD^+ with egg extracts greatly increased its Ca^{2+} releasing activity and eliminated the delay in the onset of release, indicating that it was due to enzymatic conversion of NAD^+ to an active metabolite.[6] These results set the stage for the eventual purification and identification of the active metabolite as cADPR. Comparing the Ca^{2+} releasing activity in the same egg homogenate preparation, cADPR is generally much more effective than IP_3, releasing more Ca^{2+} at seven- to tenfold lower concentrations.

Since the original demonstration that cADPR is active in releasing Ca^{2+} in sea urchin eggs, various other cells have since been found to be responsive. As listed in Table 1, they include mammalian, amphibian, and marine invertebrate cells, indicating that cADPR is likely to have general relevance.

Considerable information has also been amassed in the past several years concerning various components of the cADPR pathway, which is summarized in Figure 1A. The synthesis of cADPR from NAD^+ is catalyzed by ADP-ribosyl cyclase (C).[16] It is degraded by cADPR hydrolase (H) to ADP-ribose.[17] The cyclase activity is stimulated by a cGMP-dependent process which is likely to be mediated by a cGMP-dependent protein kinase (G) (see also Chapter 4 of this volume).[11] NO is known to raise cGMP levels in cells and could be involved in activating the pathway. The pharmacology of the Ca^{2+} release induced by cADPR is consistent with it being mediated by a RyR-like Ca^{2+} release channel (R).[18,19] Photoaffinity labeling using an analog of cADPR shows that two major receptors for cADPR are 140 kDa and 100 kDa proteins.[20] They could be

TABLE 1
Cells Responsive to Cyclic ADP-Ribose

Cell	Ref.
Sea urchin eggs[6]	Clapper, D. L., Walseth, T. F., Dargie, P. J., and Lee, H. C. (1987) *J. Biol. Chem.*, 262, 9561.
Rat pituitary cell line, GH_4C_1[6a]	Koshiyama, H., Lee, H. C., and Tashjian, A. H. (1991) *J. Biol. Chem.*, 266, 16985.
Rat dorsal root ganglion neurons[6b]	Currie, K. P., Swann, K., Galione, A., and Scott, R. H. (1992) *Mol. Biol. Cell,* 3, 1415.
Rat pancreatic islets and cerebellar microsomes[6c]	Takasawa, S., Nata, K., Yonekura, H., and Okamoto, H. (1993) *Science,* 259, 370.
Rat brain microsomes[6d]	White, A., Watson, S. P., and Galione, A. (1993) *FEBS Lett.,* 318, 259.
Cardiac sacroplasmic reticulum and brain microsomes[6e]	Meszaros, L. G., Bak, J., and Chu, A. (1993) *Nature,* 364, 76.
Pancreatic acinar cells[6f]	Thorn, P., Gerasimenko, O., and Petersen, O. H. (1994) *EMBO J.,* 13, 2038.
Bullfrog sympathetic neurons[6g]	Hua, S. Y., Tokimasa, T., Takasawa, S., Furuya, Y., Nohmi, M., Okamoto, H., and Kuba, K. (1994) *Neuron,* 12, 1073.
Guinea-pig ventricular myocytes[6h]	Rakovie, S. and Terrar, D. A. (1994) *J. Physiol.,* 475. P, 81.

accessory proteins associated with the RyR-like channel or may represent variant forms of RyR. The Ca^{2+} release activity of cADPR is strongly stimulated by calmodulin (CaM) and Ca^{2+}.[21] The action of CaM appears to be to increase the sensitivity of the release mechanism to Ca^{2+}. In the remainder of this article, each of these components of the pathway will be described in detail.

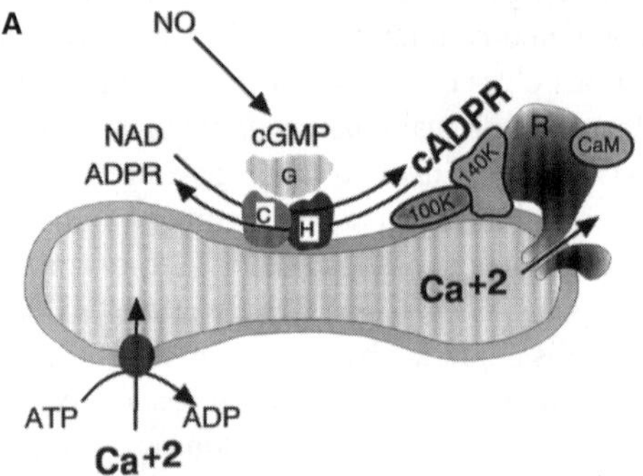

FIGURE 1. (A) Components involved in the Ca^{+2}-signaling pathway mediated by cyclic ADP-ribose. Details of the pathway are described in the text. Abbreviations used are NO, nitric oxide; ADPR, ADP-ribose; C, ADP-ribosyl cyclase; H, cADPR hydrolase; G, cGMP-dependent protein kinase; 100 K, the 100 kDa protein identified by photoaffinity labeling; 140 K, the 140 kDa protein identified by photoaffinity labeling; R, ryanodine-receptor like $Ca+^2$ channel; CaM, calmodulin.

The main unknown of the pathway is how it is activated by external ligands. This is due in part to the technical difficulty in measuring changes in the endogenous levels of cADPR. At the present time, the only available assay is the bioassay based on Ca^{2+} release activity in egg homogenates. Although it is highly specific and has been applied to measuring the natural occurrence of cADPR in tissues,[9] its sensitivity is still considerably lower than the radioimmunoassay available for other cyclic nucleotides. The low endogenous

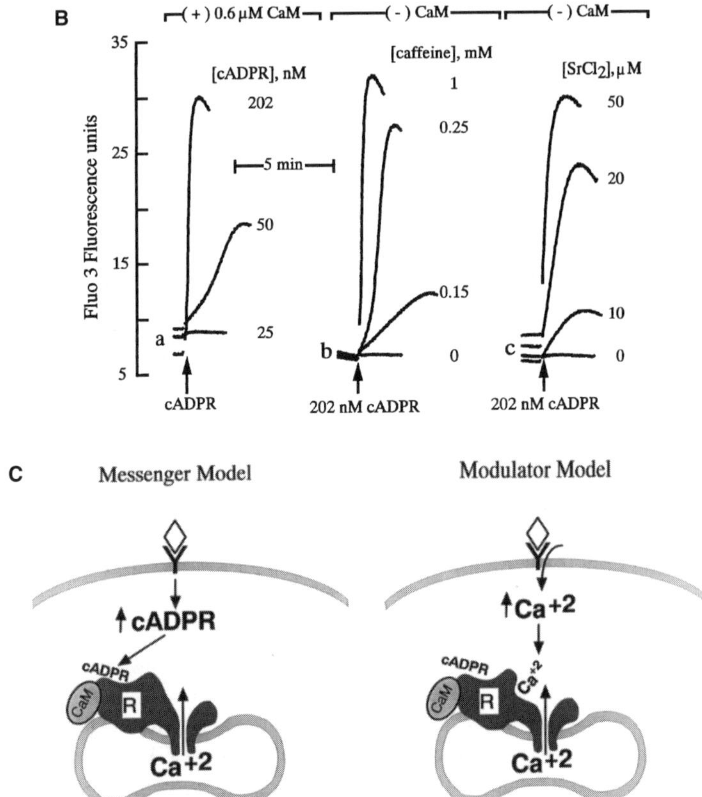

FIGURE 1 (continued). (B) Sensitization of the microsomal Ca^{2+} release system with caffeine and Sr^{2+} can substitute for CaM in conferring cADPR sensitivity. Egg microsomes (*S. purpuratus*) were incubated with 0.6 μ*M* CaM ((+)CaM, a) or without ((-)CaM, b and c); (a) Ca^{2+} release was induced by various concentrations of cADPR (25 to 202 nM); (b) caffeine (0 to 1 mM) was added to the microsomes 1 to 2 min prior to the addition of cADPR, (c) Sr^{2+} (0 to 50 μ*M*) was added to the microsomes 1 to 2 min prior to the addition of cADPR. Sr^{2+} was used instead of Ca^{2+} because it does not interfere with Fluo 3. Experimental details were described previously.[21] **(C)** Models depicting the two modes of action of cyclic ADP-ribose. The activation signal in the messenger model is the ligand-induced increase in cytosolic cADPR, while in the modulator model, it is agonist-gated influx of Ca^{2+}. In the latter case, the primary function of cADPR and calmodulin (CaM) is to sensitize the RyR-like Ca^{2+} channel (R) to Ca^{2+}.

concentrations of cADPR necessitate the use of large amounts of tissue, making it infeasible to analyze a large number of samples. It is anticipated that once the technical difficulty is solved, progress will be made in understanding the signal transduction mechanism of the pathway.

The fact that cADPR and calmodulin synergistically enhance the Ca^{2+} sensitivity of the release mechanism suggests another possible mode of action for cADPR,[21] which is to act as a modulator of CICR. In the modulator model shown in Figure 1C, cADPR is envisioned to be always present in cells at low concentrations which, together with calmodulin, is sufficient to sensitize the RyR-like channel to Ca^{2+}. The rise in intracellular Ca^{2+} concentration due to either receptor-activated Ca^{2+} influx or mobilization of internal stores via the IP_3 pathway would activate the sensitized channel and result in further Ca^{2+} release. This modulator mechanism could be involved in the propagation of Ca^{2+} waves observed in sea urchin egg during fertilization as well as in many other cells. This model is different from the standard messenger model, also shown in Figure 1C, in that the activation signal would come from the increase in intracellular Ca^{2+} concentration and not necessarily from elevation of cADPR levels. Indeed, the modulator model would predict that the endogenous levels of cADPR would be maintained at a relatively constant level independent of ligand activation.

2.1. STRUCTURE OF CYCLIC ADP-RIBOSE

The cyclic structure of cADPR was deduced by detailed characterization of the properties of the metabolite using various techniques including NMR, mass spectrometry, radioactive labeling, and HPLC.[7] The strongest evidence came from showing that cADPR can be hydrolyzed to ADP-ribose. High resolution mass spectrometry measurements of cADPR showed that its mass is exactly one water molecule smaller than its hydrolysis product, ADP-ribose. This is analogous to hydrolysis of cAMP, which produces AMP by the addition of one water molecule. The cyclic structure of cADPR has now been confirmed by X-ray crystallography, which also provides detailed information concerning the exact site of cyclization as well as conformation of various linkages.[8] Figures 2A and B show the metabolic pathway and the structure of cADPR as derived from X-ray crystallography. The site of cyclization is at N1 of the adenine and the linkage is in the β–conformation, the same as the other glycosidic linkage to the adenine (Figure 2A). Hydrolysis of the N1-glycosidic linkage by addition of a water molecule, as shown in the Figure 2A, produces ADP-ribose. The hydrolysis process occurs spontaneously with a half-time of about 24 hours at pH 2.0 and 37°C.[17] The rate decreases to about one-half at pH 7.0. The process is also catalyzed by cADPR hydrolase as described below.

The linkage between N6 and C6 of the adenine is a double bond. The molecule is thus uncharged when crystallized. In solution at physiological pH, the adenine can become protonated, with N6 converted to a primary amine group. Based on UV spectroscopic analyses, the adenine group was determined to have a pKa value of about 8.3.[22] That the adenine group is in equilibrium between a charged and uncharged state could have important implications concerning how cADPR interacts with its receptor.

A CPK-view of the molecule shows that it does not really resemble a ring but is rather compact with very little opening in the center. The C8 of the adenine is directed outward while the C2 is toward the center.[8] This orientation of the adenine ring is consistent with the fact that the hydrogen attached to C8 is amenable to modification, and a series of analogs of cADPR have been made with various substitutions attached to C8. These analogs were produced by first synthesizing the respective 8-substituted NAD^+ and then enzymatically cyclizing it with ADP-ribosyl cyclase.[23] As was observed, modifications at that position did not hinder the cyclizing reaction since C8 is directed outward. One would predict this would not be the case for modifications at the C2 position. As C2 is directed toward the center, modification at that position is likely to generate steric hindrance, prohibiting the cyclizing reaction.

2.2. METABOLIC ENZYMES

The metabolic pathway of cADPR is outlined in Figure 2A. ADP-ribosyl cyclase catalyzes the synthesis of cADPR from NAD^+ by linking the terminal ribose with the adenine at the N1 position.[16] The nicotinamide group of NAD^+ is released in the process. The enzyme is widely distributed among animal tissues.[24] Table 2 lists various tissues we have tested that have detectable cyclase activity. Even though the cyclase activity was first detected in sea urchin egg extracts, in fact, it is among the tissues which has the lowest activity. The highest activity is in the *Aplysia* ovotestis, which is a million-fold higher than any other tissues reported. The *Aplysia* enzyme is a 30 kDa soluble protein that has been purified, sequenced, and crystallized (reviewed in Reference 25).

The enzymatic reaction catalyzed by the cyclase does not require Mg^{2+} as cofactor and can partially use NADH as substrate.[16] These properties, together with the fact that the enzyme has to cyclize NAD^+, a rather long and linear molecule, make the reaction quite unique and unusual. Indeed, based on the crystal structures of cADPR and NAD^+, it can be inferred that the NAD^+ binding site of the cyclase is likely to be very different from those of many of the dehydrogenases that use NAD^+ as a coenzyme. Figure 2B shows a stereo view of NAD^+ bound to the active site of malate dehydrogenase.[26] As can be seen, the bound NAD^+ is in an extended configuration and the N1 of the adenine group is separated from the anomeric carbon of the terminal ribose by a distance of more than 12 Å. For comparison, a stereo view of cADPR is also shown with the adenine ring oriented in the same direction as the NAD^+. It is clear that, for the cyclization reaction to occur, the active site of the cyclase must be able to bind NAD^+ in a folded configuration such that the N1 of the adenine can be brought into proximity of the terminal ribose. ADP-ribosyltransferase, which catalyzes the ADP-ribosylation reaction, is another class of enzymes that uses NAD^+ as substrate. Diphtheria toxin belongs to this

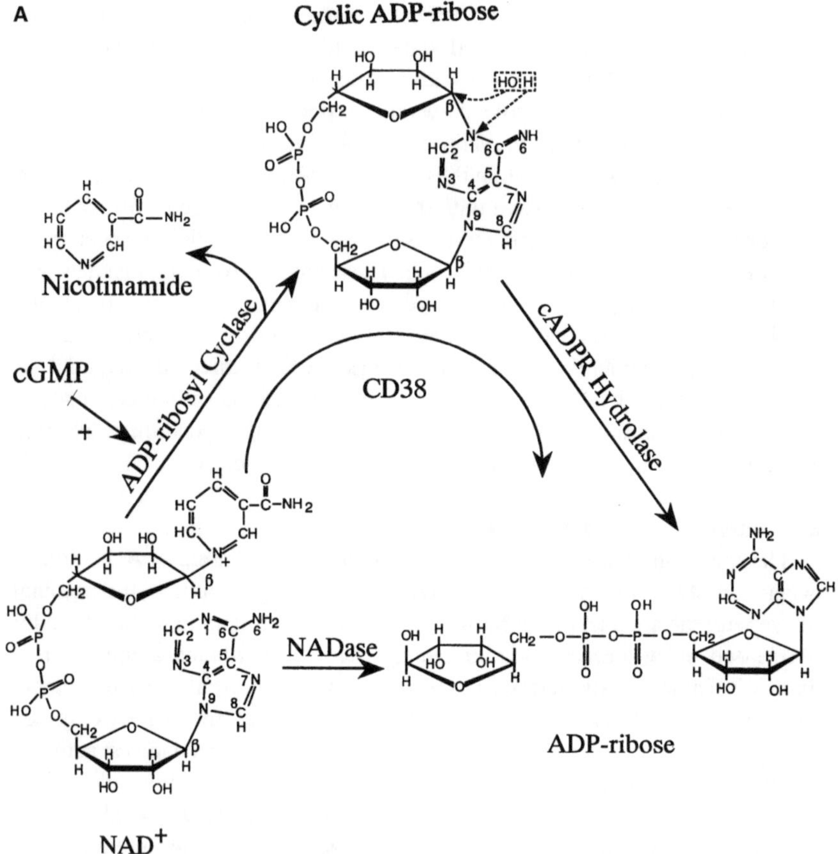

FIGURE 2. (A) Metabolic enzymes of cyclic ADP-ribose. ADP-ribosyl cyclase is stimulated (+) by a cGMP-dependent process. CD38 is a lymphocyte differentiation antigen which can catalyze both the synthesis and hydrolysis of cADPR. **(B)** Stereo views of the crystal structures of cyclic ADP-ribose and NAD^+. The two molecules are aligned with the adenine ring in the same orientation.

family and its crystal structure has been determined. The rough position of NAD^+ in the active site was inferred from the structural data obtained from the toxin complexed with a competitive analog of NAD^+, adenylyl 3′,5′-uridine monophosphate, and it appears that it is also in the extended conformation.[27] Therefore, it is likely that the active site of the cyclase is rather unique. Now that the *Aplysia* cyclase has been crystallized,[25] its crystal structure should provide important information concerning the conformation of the active site as well as insights to its novel cyclizing activity.

As shown in the messenger model of Figure 1C, a possible mode of action of cADPR is to serve as a second messenger. This model requires that stimuli, such as binding of ligands to surface receptors, would increase the synthesis of

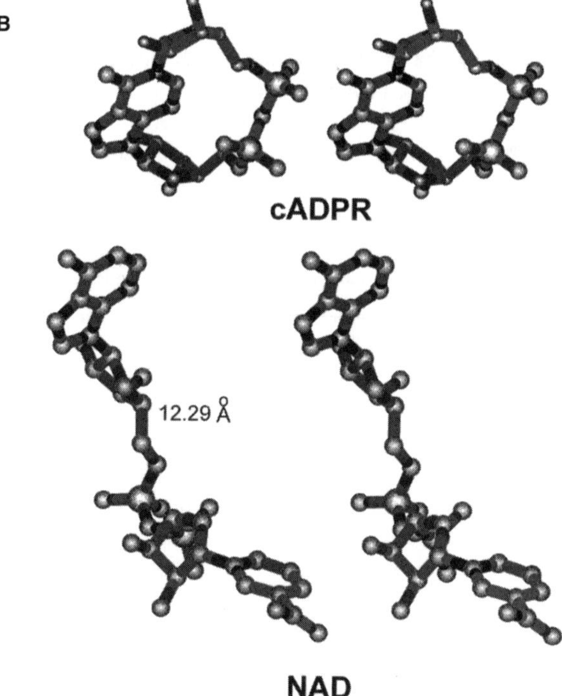

FIGURE 2. (continued)

cADPR. The recent finding that the cyclase activity in sea urchin egg extracts can be stimulated by a cGMP-dependent process is consistent with such a scheme.[11] The cGMP-dependent activation of the cyclase was found to be inhibitable by protein kinase inhibitors, suggesting that protein phosphorylation may be involved in regulating the cyclase activity (see also Chapter 4 of this volume). Although the sea urchin cyclase has not been purified and its amino acid sequence is unknown, examination of the sequence of the *Aplysia* cyclase shows that there are numerous consensus phosphorylation sites for cGMP-dependent kinase (reviewed in Reference 25). That cGMP can modulate the cyclase activity suggests that NO may be able to activate the cADPR pathway by elevating cellular cGMP levels. So far, NO has been shown to mobilize intracellular Ca^{2+} in two cell types[28,29] and in one of these cells, the interstitial cells of the colon, this effect of NO can be inhibited by ryanodine.[28] As will be discussed below, accumulating evidence strongly implicates cADPR as an endogenous modulator of a RyR-like Ca^{2+} channel, making the connection between the NO and the cADPR pathways a very likely possibility.[30]

The degradation pathway of cADPR consists of cADPR hydrolase catalyzing the hydrolysis of the N1-glycosidic linkage to produce ADP-ribose (Figure 2A). The hydrolase activity is equally as widespread as the cyclase and has been detected in mammalian, avian, amphibian, and marine invertebrate

TABLE 2
Widespread Occurrence of ADP-Ribosyl Cyclase in Animal Tissues

Species	Tissues
Mammal	
Human	Red blood cell, platelet, myeloma cell
Sheep	Trachea smooth muscle
Dog	Brain, liver, aorta, carotid, testis
Rabbit	Brain, liver, heart, kidney, spleen, retina
Rat	Brain, liver, heart
Avian	
Chick	Embryonic brain
Amphibian	
Salamander	Brain
Insect	
Drosophila	Embryo
Marine Invertebrate	
Sea Urchin	Egg
Aplysia	Ovotestis

tissue extracts.[17] The activity in mammalian brain tissues is particularly high and is the source from which it has been partially characterized. It is an membrane-bound enzyme with a K_m of about 0.16 m*M* and a broad pH dependency with an optimum around neutrality. The brain enzyme has not been purified.

In addition to the cyclase and hydrolase, there is a new class of bifunctional enzymes which is also involved in the metabolism of cADPR. These bifunctional enzymes not only can synthesize cADPR from NAD^+, but also can hydrolyze cADPR to ADP-ribose (Figure 2A). The discovery of this novel class of enzymes raises the possibility that the cADPR hydrolase activity detected in tissue extracts may not be catalyzed by a monofunctional enzyme but, instead, may represent the hydrolytic part of the reaction catalyzed by these bifunctional enzymes. The first member of this family is an enzyme purified from spleen which is a membrane-bound protein of 39 kDa.[31] The cyclizing reaction, from NAD^+ to cADPR, catalyzed by the spleen enzyme was found to be reversible since incubation of the enzyme with cADPR and nicotinamide resulted in formation of NAD^+. Bifunctional enzymes are not common and the novel catalytic properties of the bifunctional enzymes involved in the metabolism of cADPR make them especially interesting.

Another member of this enzyme class is CD38, a lymphocyte differentiation antigen.[32] It turns out that the amino acid sequence of the *Aplysia* cyclase has considerable homology with CD38.[33] The two proteins not only have 30% identity and 40% conservative substitutions, but also the perfect alignment of the positions of ten of the cysteine residues. Based on these similarities, it was postulated that CD38 may have cyclase activity.[33] Indeed, a recombinant CD38 has been shown to have both cADPR synthesizing as well as hydrolyzing activities.[32] Similarly, native CD38 purified from the surface of human red blood cells also exhibits bifunctionality.[34] This novel class of bifunctional enzymes may turn out to be an extended family. Indeed, a glycosyl-phosphatidylinositol-anchored protein called BST-1 was found to have 33% sequence identity with CD38 and 26% with *Aplysia* cyclase.[35] In addition, nine of the ten cysteine residues in the molecule are also perfectly aligned with those of the cyclase. BST-1 has been suggested to be involved in facilitation of pre-B-cell growth. Whether BST-1 indeed has enzyme activity has not been demonstrated.

So far, it appears that the two best-characterized metabolic enzymes of cADPR, the *Aplysia* cyclase and CD38, are not likely to be involved in regulating cellular cADPR levels. This is because the *Aplysia* cyclase is found inside cytoplasmic granules[36] while CD38 is an ectoenzyme.[32,34] Their peculiar locations in cells preclude them from being directly involved in the Ca^{2+}-signaling pathway. It is more than likely that there is yet another form of the enzyme present in cells. A prime candidate for the signaling cyclase is the one detected in sea urchin egg homogenates which is sensitive to cGMP[11] (see also Chapter 4 of this volume). Unlike CD38, this enzyme is intracellular since no cyclase activity can be measured after incubation of intact eggs with NAD^+.[47] Also, its sensitivity to cGMP certainly implicates its role in the signaling process.

In tissue extracts where both the cyclase and hydrolase activities are present, the overall reaction is the conversion of NAD^+ to ADP-ribose, which is the same as that catalyzed by the classical NADases (Figure 2A). This is also the case for the CD38-like enzymes. Indeed, during the reaction catalyzed by these enzymes, the steady-state concentration of cADPR, the balance between the cyclization and hydrolysis reaction, is generally very low as compared with ADP-ribose, the final product, and can be easily missed.[31,32,34] This raises the possibility that all enzymes previously classified as NADases may, in fact, be the metabolic enzymes of cADPR. This, as it turns out, is not the case. A classical NADase from *Neurospora* (Sigma) catalyzes efficient hydrolysis of NAD^+ to ADP-ribose. It produces no detectable cADPR nor does it catalyze its hydrolysis.[48] It is truly an NADase as originally classified. However, another enzyme purified from snake venom, which is also classified as NADase, turns out to be a bifunctional enzyme like CD38.[31] It is thus clear that NADase is not a homogeneous class of enzymes. Because cADPR was not known prior to 1987, some of the enzymes previously classified as NADase, in fact, may be metabolic enzymes of cADPR. Therefore, it is pertinent that the classification of NADases be carefully reexamined, taking into consideration the current knowledge of cADPR.

2.3. RELATIONSHIP WITH CALCIUM-INDUCED CALCIUM RELEASE

Table 3 compares the pharmacology between the cADPR and the IP_3 pathways in sea urchin egg microsomes. The half-maximal concentration of cADPR for inducing Ca^{2+} release is about 18 n*M* as determined in *Lytechinus*

TABLE 3
Pharmacology of the cADPR- and IP_3-Dependent Calcium Release

	cADPR	IP_3
Half-maximal concentration	18 v*M* K_d = 17 n*M*	130 n*M*
Specific binding of ^{32}-P-cADPR	Not affected by heparin or IP_3	—
High concentrations of IP3	No effect	Desensitized
High concentrations of cADPR	Desensitized	No effect
Antagonist of IP_3		
Heparin	No effect	Inhibited
Antagonists of cADPR		
8-amino-cADPR	Inhibited	No effect
8-Br-cADPR	Inhibited	No effect
8-azido-cADPR	Inhibited	No effect
Antagonists of Ca^{+2}-induced Ca^{+2} release		
Procaine	Inhibited	No effect
Ruthenium red	Inhibited	No effect
Agonists of Ca^{+2}-induced Ca^{+2} Release		
Low concentrations of caffeine	Potentiated	No effect
Low concentrations of Ca^{+2} or Sr^{+2}	Potentiated	No effect
High concentrations of caffeine	Desensitized	No effect
High concentrations of ryanodine	Desensitized	No effect

pictus microsomes.[37] In the same preparation, the half-maximal concentration for IP_3 was found to be seven to ten times higher, indicating that cADPR is much more effective.[37] That cADPR is active in the nanomolar range suggests that it may operate through a specific receptor. This was demonstrated using ^{32}P-cADPR synthesized by incubation of ^{32}P-NAD$^+$ with ADP-ribosyl cyclase (Table 3).[15] Specific binding of ^{32}P-cADPR to *Strongylocentrotus purpuratus* microsomes was detected.[15] The binding is saturable and competitively inhibited by nanomolar concentrations of unlabeled cADPR. The dissociation constant determined by Schatchard analysis was about 17 n*M* and the receptor density was about 25 fmol/mg of microsomal protein.[15] The binding was unaffected by IP_3 up to 10 μ*M* and was not sensitive to inhibition by heparin, indicating that the cADPR receptor is different from the IP_3 receptor.[15] More recent results show that there is a heat labile and low molecular weight factor present in the egg homogenates which inhibits specific binding of cADPR to egg microsomes. [25] Removal of the inhibitor by dialysis unmasked two type of binding sites in *S. purpuratus* microsomes, a high affinity site with a Kd about 0.7 n*M* and a low affinity site with a K_d of about 170 n*M*.[25]

The K_d of the low affinity binding site is close to the concentration of cADPR which produces half-maximal Ca^{2+} release measured in the same species of egg microsomes,[19] suggesting that the low affinity site may be the active site while the high affinity site may be inactive. The presence of two types of binding sites in the microsomes may explain the desensitization of the cADPR-mediated Ca^{2+} release. Thus, binding of cADPR to the low affinity sites may activate Ca^{2+} release and convert the sites to high affinity, which then become inactive and, therefore, desensitized. A similar mechanism has been proposed to account for the desensitization of the IP_3-mediated Ca^{2+} release in liver microsomes.[38] On the other hand, in the case of sea urchin egg microsomes, a possibility that needs consideration is that the desensitization may be due to discharge of the cADPR-sensitive stores that are separate from the IP_3-sensitive stores. This possibility, however, is not consistent with the observation that the cADPR- and IP_3-induced Ca^{2+} release were not additive.[37]

Irrespective of the exact mechanism of desensitization, it does provide a convenient way to demonstrate the independence between the cADPR- and the IP_3-sensitive release (Table 3). Thus high concentrations of either cADPR or IP_3 can desensitize the egg microsomes only to the same agonist but not to the other.[37] Heparin, although not effective in blocking cADPR binding to its microsomal receptor, is very potent in inhibiting the IP_3-sensitive Ca^{2+} release without affecting the cADPR-dependent component.[37] Conversely, 8-amino-cADPR, a novel inhibitor of the cADPR action, at nanomolar concentrations effectively blocks only the cADPR-sensitive release but not the IP_3-sensitive release.[23,39] Other 8-substituted analogs of cADPR such as the 8-Br- and the 8-azido-derivatives are all antagonists of the cADPR-dependent mechanisms with varying degree of effectiveness, and none of them affects the Ca^{2+} release induced by IP_3.[23] That the cADPR pathway is separate and independent from the IP_3 pathway has also been demonstrated in intact sea urchin eggs by microinjection of 8-amino-cADPR and/or heparin. Blocking each of the pathways individually is not sufficient to inhibit the Ca^{2+} mobilization associated with fertilization. Both pathways needed to be suppressed before total inhibition of the Ca^{2+} changes could be achieved.[39]

In addition to the IP_3 pathway, CICR is another major mechanism for mobilizing intracellular Ca^{2+}. Table 3 shows that antagonists of CICR, such as ruthenium red and procaine, were found to inhibit the cADPR-induced Ca^{2+} release without affecting the IP_3-sensitive release.[18] Agonists of CICR, such as caffeine and divalent cations at low concentrations greatly potentiate the cADPR-sensitive Ca^{2+} release.[19] Furthermore, at high concentrations caffeine can selectively desensitize egg microsomes to cADPR without affecting their response to IP_3.[18,19] Similarly, high concentrations of ryanodine, a modulator of CICR, can also desensitize the microsomal response to cADPR selectively.[18] Taken together, these characteristics of the cADPR-sensitive Ca^{2+} release indicate it is independent from the IP_3 pathway and its pharmacology is consistent with it being mediated by a CICR mechanism, which is likely to be a RyR-like Ca^{2+} channel.

Specific cADPR-binding proteins have been identified in egg microsomes using 8-azido-cADPR as a photoaffinity labeling agent.[20] This analog is an antagonist of the cADPR-induced Ca^{2+} release and it specifically competes for the microsomal cADPR binding sites. The two cADPR receptor proteins identified by this technique are proteins of 140 kDa and 100 kDa, which are smaller than the 380 kDa protein that cross reacts on Western blots with an antibody against the skeletal RyR.[40] However, the sizes of the 140 kDa and the 380 kDa proteins are consistent with the fragments of the RyR cleaved by endogenous calpain-like protease.[41] It is possible that in sea urchin eggs the RyR-like receptors are processed posttranscriptionally to smaller proteins. In view of the very specialized requirement for the large cytoplasmic foot structure of the RyR in excitation-contraction coupling in muscle, the proteolytic processing of the RyR-like receptor in sea urchin egg is very likely, since there is no special need for such an electromechanical coupling between the plasma membrane and the endoplasmic reticulum in the egg. The facts that high concentrations of protease inhibitors are always present during preparation of egg microsomes and that the cADPR receptor is fully functional during the photoaffinity labeling,[20] indicate that proteolytic processing, if it occurs, may indeed be a normal posttranscriptional modification.

Alternatively, the 140 kDa and the 100 kDa proteins may represent accessory proteins of the RyR. It is now known that several proteins are tightly associated with the RyR. For example, FKPB12, a binding protein for the immunosuppressor drug FK506, was found to be associated with and co-purified with the RyR through multiple steps of chromatography.[42] Likewise, 5 or 6 mol of calmodulin were shown to be tightly bound to each subunit of the skeletal RyR with a dissociation constant of about 8 n*M* even in the absence of Ca^{2+}.[43] Indeed, as will be described in detail below, the cADPR-sensitive Ca^{2+} release has a near-absolute dependence on calmodulin.[21]

2.4. REGULATION BY CALMODULIN

In the original paper describing the Ca^{2+} mobilizing effects of cADPR, it was stated that the activity requires a soluble protein factor.[6] This was demonstrated by purifying the Ca^{2+} release microsomes using Percoll density centrifugation. After centrifugation, the homogenates can be fractionated into two main regions on the gradient. The bottom of the tube contains the majority of the organelles which include mitochondria, yolk granules, and other denser organelles. The microsomes that response to cADPR and IP_3 are concentrated in a tight band on the upper part of the gradient.[6,14,15] The microsomes can be collected without contamination by inserting a needle into the band. The soluble proteins can also be collected from the supernatant of the gradient.

The purified microsomes do not release Ca^{2+} even when challenged with a saturating concentration of cADPR.[21] The responsiveness can be restored by reconstituting the system with soluble proteins in the supernatant. The restoration of the cADPR reactivity shows concentration-dependence on the supernatant proteins.[21] These results indicate the presence of a soluble protein in the supernatant mediating the cADPR response. This cADPR sensitivity-conferring factor turns out to be highly conserved since it is also present in soluble extracts prepared from dog brains.[21] Using the cADPR responsiveness of the egg microsomes as a bioassay, the soluble factor has been purified to apparent homogeneity from brain and sea urchin egg extracts using various chromatography steps. Both the egg factor and the brain factor are proteins of about 17 kDa. They both are heat stable, retaining full activity in boiling water for 5 to 10 min. Their mobility on SDS-gel is Ca^{2+}-dependent, indicating they are Ca^{2+} binding proteins. Finally, they both have very low absorbance at 280 nm, indicating they have low tryptophan content. These properties are consistent with the factor being calmodulin. This was verified by Western analyses of the purified factor using a monoclonal antibody against calmodulin.[21] The antibody shows similar reactivity to both the brain and the egg factor, consistent with the factor being calmodulin. The antibody can also detect the presence of calmodulin in the crude supernatant of the sea urchin egg. Furthermore, both the egg and the brain factor were found to be as effective as authentic calmodulin in stimulating bovine heart phosphodiesterase.[21] Perhaps the most direct proof of the factor being calmodulin is that authentic calmodulin can substitute for the factor in conferring cADPR sensitivity to the microsomes. Indeed, it was shown that calmodulin not only can confer cADPR sensitivity to the microsomes but it can do so with the same effectiveness as the brain and egg factors.[21]

The dependence of microsomal cADPR-sensitivity on calmodulin is remarkably strong. In the absence of calmodulin, the microsomes would not release Ca^{2+} even when challenged with saturating concentrations of cADPR, while concentrations as low as the nanomolar range would give significant Ca^{2+} release in the presence of calmodulin.[21] Therefore, depending on how much calmodulin is present, the responsiveness of the microsomes to cADPR can vary by two to three orders of magnitude. Although the cADPR-dependent Ca^{2+} release shows critical dependence on calmodulin, it is not the case for the IP_3-dependent pathway. Whether calmodulin is present or not, the egg microsomes show the same concentration dependency to IP_3. A calmodulin antagonist, W7, can also inhibit the cADPR-sensitive Ca^{2+} release in crude egg homogenates selectively without affecting the microsomal response to IP_3.[21] The requirement for calmodulin is therefore selective for the cADPR pathway.

The mechanism of the calmodulin action does not appear to involve enzymatic reactions, since various inhibitors of calmodulin-dependent phosphatase and kinase were found to have no effect on the cADPR-conferring activity of calmodulin.[21] Instead, calmodulin appears to sensitize the CICR system in the microsomes. As shown in Figure 1B, two other agonists of CICR, caffeine and Sr^{2+}, can also confer cADPR sensitivity to the microsomes. In the presence of calmodulin, 202 nM of cADPR induced maximal Ca^{2+} release (a in Figure 1B). In the absence of calmodulin, the same concentration of cADPR produced no Ca^{2+} release (b in Figure 1B). However, in the presence of 1 m*M* caffeine, addition of cADPR produced as much Ca^{2+} release as in the presence of calmodulin. Even as low as 0.15 m*M* caffeine can confer significant cADPR-sensitivity to the microsomes. Indeed, 1 mM caffeine, in fact, represents a saturating concentration as far as conferring cADPR sensitivity is concerned.

It is generally believed that caffeine acts by increasing the Ca^{2+} sensitivity of the CICR. Therefore, elevating the divalent concentration should have the same effect as caffeine. This is shown in c of Figure 1B. In the presence of 50 μM Sr^{2+}, cADPR can produce as much Ca^{2+} release as in the presence of calmodulin. Another way of interpretating these results is that the Ca^{2+} sensitivity of the release mechanism is increased by the presence of cADPR. It has previously been shown that in the absence of cADPR, Sr^{2+} alone is not sufficient to produce any Ca^{2+} release unless it is in the range of 100 μM.[19] In the presence of cADPR, the release mechanism is sensitized to divalent cations, and even low concentrations of Sr^{2+} can now induce a large Ca^{2+} release as shown in Figure 1B, graph c. The similarity of the effects of calmodulin, caffeine, and cADPR suggests all three agonists are synergistic in sensitizing the Ca^{2+} release system to Ca^{2+}. Indeed, in the presence of saturating concentrations of both calmodulin and cADPR, the release system is so sensitized that even the ambient Ca^{2+}, which has been measured to be in the range of 20 to 50 nM,[19] appears to be sufficient to trigger Ca^{2+} release. In this view, therefore, the activating signal is not cADPR but the ambient Ca^{2+}; cADPR and calmodulin are mainly acting as sensitizers. This interpretation, of course, is the basis for the modulator model described in Figure 1C.

The fact that totally unrelated chemicals such as caffeine and Sr^{2+} can substitute for calmodulin in conferring cADPR sensitivity to microsomes suggests that calmodulin acts by direct interaction with the Ca^{2+} release mechanism, most likely the RyR-like Ca^{2+} channel. This mode of action is similar to that observed in the cyclic nucleotide-activated channel of rat olfactory receptor neurons.[44] In that case, the sensitivities of the cation channel for cAMP and cGMP were changed 10- to 20-fold in the presence of calmodulin and the modulating effects of calmodulin require Ca^{2+}. In the case of the cADPR-dependent Ca^{2+} channel, the sensitizing effect of calmodulin appears to be independent of Ca^{2+}, since it was observed when the ambient Ca^{2+} concentration was in the nanomolar range. This is consistent with the previous demonstration by photoaffinity labeling that calmodulin binds specifically to RyR in the absence of Ca^{2+}.[43]

Direct modulation by calmodulin on cardiac and skeletal RyR has been observed with single-channel recordings.[45] The effects appear to be inhibitory and show specific requirements for the presence of Ca^{2+} as well as the absence of ATP. The sensitivity of the inhibitory effects to experimental variables make it difficult to compare them directly with the sensitizing effects of calmodulin seen with the egg microsomes because the conditions are very different. It is perhaps much more pertinent to determine if the single-channel activity of RyR shows higher sensitivity toward cADPR in the presence of calmodulin, which as yet remains to be measured. The situation could be much more complicated if the cADPR action is mediated by accessory proteins, e.g., the 140 kDa and the 100 kDa proteins identified by photoaffinity labeling, working in concert with calmodulin. In any case, the identification of the cADPR binding proteins or receptors is of crucial importance in clarifying the details of the molecular machinery involved in the action of cADPR. Now that they are successfully labeled, it should set the stage for their eventual identification.

3. SUMMARY

The cADPR pathway was discovered in 1987 because of a curious observation[6] and has since been shown to be mediated by a hitherto unknown metabolite of NAD^+.[7,8] The novelty of these results, together with the fact that the first cell, the sea urchin egg, shown to be responsive to cADPR is an invertebrate cell of limited appeal, brought forth the suspicion that the pathway might not have any general relevance. This has now been shown to be incorrect, as both the cellular responsiveness to cADPR and its metabolic enzymes have been found to be widespread among various species of animals and cell types. Persistent investigations of its mechanism of action in sea urchin eggs have provided much of our present understanding of the pathway. It is now clear that the cADPR pathway is emerging as one of the most versatile signaling mechanisms. Indeed, its multiple connection points with other established pathways suggest it may function as an integrative mechanism. The pathway is linked to cyclic nucleotide signaling through the stimulating effect of cGMP on the ADP-ribosyl cyclase. This, in turn, could provide a means for NO to mobilize internal Ca^{2+}. As shown in the standard messenger model of Figure 1C, the cADPR pathway may also be involved in transducing surface receptor activation linked to elevation of cellular cGMP. Through activation of ADP-ribosyl cyclase, the cGMP increase can lead to mobilization of internal stores. Alternatively, as depicted in the modulator model (Figure 1C), the ability of cADPR and calmodulin to increase the sensitivity of the Ca^{2+} release mechanism to Ca^{2+} furnishes a mechanism for amplifying as well as propagating the effects of Ca^{2+} elevation produced by agonist-gated Ca^{2+} influx and/or IP_3-induced Ca^{2+} mobilization. In addition to these positive controls, the cADPR pathway is also self-limiting by desensitization, which can function as a negative feedback mechanism.

Perhaps the most intriguing aspect of the cADPR pathway is the discovery that surface antigens such as CD38 are bifunctional enzymes participating in the metabolism of cADPR. This raises the possibility that cADPR may have extracellular functions as well. This would be analogous to cAMP acting as a chemoattractant for *Dictyostelium*.[46] It is intriguing to speculate that cADPR may actually be a hormone with paracrine and/or endocrine functions. The availability of the photoaffinity agent, 8-azido-cADPR, should provide a straightforward means to identify the putative surface receptor for cADPR. Similarly, the physiological roles of the pathway can be readily assessed by employing the series of antagonists of cADPR, such as 8-amino-cADPR, as has been done in sea urchin eggs.[39] These technological advances should facilitate our understanding of the pathway in other cellular systems as well as discovering other aspects of the pathway in addition to its Ca^{2+} mobilizing activity. As it stands now, we may be seeing only the tip of an iceberg.

REFERENCES

1. **Rall, T. W. and Sutherland, E. W.,** Formation of a cyclic adenine ribonucleotide by tissue particles, *J. Biol. Chem.,* 232, 1065, 1958.
2. **Fesenko, E. E., Kolesnikov, S. S., and Lyubarsky, A. L.,** Induction by cyclic GMP of cationic conductance in plasma membrane of retinal rod outer segment, *Nature,* 313, 310, 1985.
3. **Beavo, J. A., Hardman, J. G., and Sutherland, E. W.,** Stimulation of adenosine 3′,5′-monophosphate hydrolysis by guanosine 3′,5′-monophosphate, *J. Biol. Chem.,* 246, 3841, 1971.
4. **Berridge, M. J.,** Inositol trisphosphate and calcium signalling, *Nature,* 361, 315, 1993.
5. **Takeshima, T., Nishimura, S., Matsumoto, T., Ishida, H., Kangawa, K., Minamino, N., Matsuo, H., Ueda, M., Hanaoka, M., Hirose, T., and Numa, S.,** Primary structure and expression from complementary DNA of skeletal muscle ryanodine receptor, *Nature,* 339, 439, 1989.
6. **Clapper, D. L., Walseth, T. F., Dargie, P. J., and Lee, H. C.,** Pyridine nucleotide metabolites stimulate calcium release from sea urchin egg microsomes desensitized to inositol trisphosphate, *J. Biol. Chem.,* 262, 9561, 1987.

6a. **Koshiyama, H., Lee, H. C., and Tashjian, A. H.,** Novel mechanism of intracellular calcium release in pituitary cells, *J. Biol. Chem.,* 266, 16985, 1991.

6b. **Currie, K. P., Swann, K., Galione, A., and Scott, R. H.,** Activation of Ca^{2+}-dependent currents in cultured rat dorsal root ganglion neurones by a sperm factor and cyclic ADP-ribose, *Mol. Biol. Cell,* 3, 1415, 1992.

6c. **Takasawa, S., Nata, K., Yonekura, H., and Okamoto, H.,** Cyclic ADP-ribose in insulin secretion from pancreatic β Cells, *Science,* 259, 370, 1993.

6d. **White, A., Watson, S. P., and Galione, A.,** Cyclic ADP-ribose-induced Ca^{2+} release from rat brain microsomes, *FEBS Letters,* 318, 259, 1993.

6e. **Meszaros, L. G., Bak, J., and Chu, A.,** Cyclic ADP-ribose as an endogenous regulator of the non-skeletal type ryanodine receptor Ca^{2+} channel, *Nature,* 364, 76, 1993.

6f. **Thorn, P., Gerasimenko, O., and Petersen, O. H.,** Cyclic ADP-ribose regulation of ryanodine receptors involved in agonist evoked cytosolic Ca^{2+} oscillations in pancreatic acinar cells, *EMBO J.,* 13, 2038, 1994.

6g. **Hua, S. Y., Tokimasa, T., Takasawa, S., Furuya, Y., Nohmi, M., Okamoto, H., and Kuba, K.,** Cyclic ADP-ribose modulates release channels for activation by physiological Ca^{2+} entry in bullfrog sympathetic neurons, *Neuron,* 12, 1073, 1994.

6h. **Rakovie, S. and Terrar, D. A.,** Possible effects of cADP-ribose on calcium transients and contractions in guinea-pig isolated ventricular myocytes, *J. Physiol.,* 475.P, 81, 1994.

7. **Lee, H. C., Walseth, T. F., Bratt, G. T., Hayes, R. N., and Clapper, D. L.,** Structural determination of a cyclic metabolite of NAD^+ with intracellular Ca^{2+} mobilizing activity, *J. Biol. Chem.,* 264, 1608, 1989.
8. **Lee, H. C., Aarhus, R., and Levitt, D.,** The crystal structure of cyclic ADP-ribose, *Nature Struct. Biol.,* 1, 143, 1994.
9. **Walseth, T. F., Aarhus, R., Zeleznikar, R. J., Jr., and Lee, H. C.,** Determination of endogenous levels of cyclic ADP-ribose in rat tissues, *Biochim. Biophys. Acta,* 1094, 113, 1991.

10. **Nakada, S., Rhee, S. K., Hamanaka, H., and Mikoshiba, K.,** Cyclic AMP-dependent phosphorylation of an immunoaffinity-purified homotetrameric inositol 1,4,5-trisphosphate receptor (type I) increases Ca^{2+} flux in reconstituted lipid vesicles, *J. Biol. Chem.,* 269, 6735, 1994.
11. **Galione, A., White, A., Willmott, N., Turner, M., Potter, B. V. L., and Watson, S. P.,** cGMP mobilizes intracellular Ca^{2+} in sea urchin eggs by stimulating cyclic ADP-ribose synthesis, *Nature,* 365, 456, 1993.
12. **Moncada, S., Palmer, R. M. J., and Higgs, E. A.,** Nitric oxide: Physiology, pathophysiology and pharmacology, *Pharmacol. Rev.,* 43, 109, 1991.
13. **Lowenstein, C. J. and Snyder, S. H.,** Nitric oxide, a novel biologic messenger, *Cell,* 70, 705, 1992.
14. **Clapper, D. L. and Lee, H. C.,** Inositol trisphosphate induces Ca^{2+}-release from non-mitochondrial stores in sea urchin egg homogenates, *J. Biol. Chem.,* 260, 13947, 1985.
15. **Lee, H. C.,** Specific binding of cyclic ADP-ribose to calcium-storing microsomes from sea urchin eggs, *J. Biol. Chem.,* 266, 2276, 1991.
16. **Lee, H. C. and Aarhus, R.,** ADP-ribosyl cyclase: An enzyme that cyclizes NAD^+ into a calcium-mobilizing metabolite, *Cell Regulation,* 2, 203, 1991.
17. **Lee, H. C. and Aarhus, R.,** Wide distribution of an enzyme that catalyzes the hydrolysis of cyclic ADP-ribose, *Biochim. Biophys. Acta,* 1164, 68, 1993.
18. **Galione, A., Lee, H. C., and Busa, W. B.,** Ca^{2+}-induced Ca^{2+} release in sea urchin egg homogenates and its modulation by cyclic ADP-ribose, *Science,* 253, 1143, 1991.
19. **Lee, H. C.,** Potentiation of calcium- and caffeine-induced calcium release by cyclic ADP-ribose, *J. Biol. Chem.,* 268, 293, 1993.
20. **Walseth, T. F., Aarhus, R., Kerr, J. A., and Lee, H. C.,** Identification of cyclic ADP-ribose binding proteins by photoaffinity labeling, *J. Biol. Chem.,* 268, 26686, 1993.
21. **Lee, H. C., Aarhus, R., Graeff, R., Gurnack, M. E ., and Walseth, T. F.,** Cyclic ADP-ribose activation of the ryanodine receptor is mediated by calmodulin, *Nature,* 370, 307, 1994.
22. **Kim, H., Jacobson, E. L., and Jacobson, M. K.,** Position of cyclization in cyclic ADP-ribose, *Biochem. Biophys. Res. Commun.,* 194, 1143, 1993.
23. **Walseth, T. F. and Lee, H. C.,** Synthesis and characterization of antagonists of cyclic ADP-ribose, *Biochim. Biophys. Acta,* 1178, 235, 1993.
24. **Rusinko, N. and Lee, H. C.,** Widespread occurrence in animal tissues of an enzyme catalyzing the conversion of NAD^+ into a cyclic metabolite with intracellular Ca^{2+} mobilizing activity, *J. Biol. Chem.,* 264, 11725, 1989.
25. **Lee, H. C., Galione, A., and Walseth, T. F.,** Cyclic ADP-ribose: Metabolism and calcium mobilizing function, in *Vitamins and Hormones,* Litwack, G., Ed., Academic Press, Orlando, FL, 48, chap. 5.
26. **Hall, M. D. and Banaszak, L. J.,** Crystal structure of a ternary complex of *Escherichia coli* malate dehydrogenase citrate and NAD at 1.9 Å resolution, *J. Mol. Biol.,* 232, 213, 1993.
27. **Choe, S., Bennett, M. J., Fujii, G., Curmi, P. M. G., Kantardjieff, K. A., Collier, R. J., and Eisenberg, D.,** The crystal structure of diphtheria toxin, *Nature,* 357, 216, 1992.
28. **Publicover, N. G., Hammond, E. M., and Sanders, K. M.,** Amplification of nitric oxide signaling by interstitial cells isolated from canine colon, *Proc. Natl. Acad. Sci. U.S.A.,* 90, 2087, 1993.
29. **Kong, S. K. and Lee, C. Y.,** The nitric oxide donor, sodium nitroprusside, increased intracellular and cytosolic free calcium concentration in single PU5-1.8 cells, *Biochem. Biophys. Res. Commun.,* 199, 234, 1994.
30. **Lee, H. C.,** A signaling pathway involving cyclic ADP-ribose, cyclic GMP and nitric oxide, *News Physiol. Sci.,* 9, 134, 1994.
31. **Kim, H., Jacobson, E. L., and Jacobson, M. K.,** Synthesis and degradation of cyclic ADP-ribose by NAD glycohydrolases, *Science,* 261, 1330, 1993.

32. **Howard, M., Grimaldi, J. C., Bazan, J. F., Lund, F. E., Santos-Argumedo, L., Parkhouse, R. M. E., Walseth, T. F., and Lee, H. C.,** Formation and hydrolysis of cyclic ADP-ribose catalyzed by lymphocyte antigen CD38, *Science,* 262, 1056, 1993.
33. **States, D. J., Walseth, T. F., and Lee, H. C.,** Similarities in amino acid sequences of *Aplysia* ADP-ribosyl cyclase and human lymphocyte antigen CD38, *Trends Biochem. Sci.,* 17, 495, 1992.
34. **Zocchi, E., Franco, L., Guida, L., Benatti, U., Bargellesi, A., Malavasi, F., Lee, H. C., and De Flora, A.,** A single protein immunologically identified as CD38 displays NAD^+ glycohydrolase, ADP-ribosyl cyclase and cyclic ADP-ribose hydrolase activities at the outer surface of human erythrocytes, *Biochem. Biophys. Res. Commun.,* 196, 1459, 1993.
35. **Kaisho, T., Ishikawa, J., Oritani, K., Inazawa, J., Tomizawa, H., Muraoka, O., Ochi, T., and Hirano, T.,** BST-1, a surface molecule of bone marrow stromal cell lines that facilitates pre-B-cell growth, *Proc. Natl. Acad. Sci. U.S.A.,* 91, 5325, 1994.
36. **Glick, D. L., Hellmich, M. R., Beushausen S., Tempst, P., Bayley H., and Strumwasser, F.,** Primary structure of a molluscan egg-specific NADase, a second-messenger enzyme, *Cell Regulation,* 2, 211, 1991.
37. **Dargie, P. J., Agre, M. C., and Lee, H. C.,** Comparison of Ca^{2+} mobilizing activities of cyclic ADP-ribose and inositol trisphosphate, *Cell Regulation,* 1, 279, 1990.
38. **Mauger, J.-P., Claret, M., Pietri, F., and Hilly, M.,** Hormonal regulation of inositol 1,4,5-trisphosphate receptor in rat liver, *J. Biol. Chem.,* 264, 8821, 1989.
39. **Lee, H. C., Aarhus, R., and Walseth, T. F.,** Calcium mobilization by dual receptors during fertilization of sea urchin eggs, *Science,* 261, 352, 1993.
40. **McPherson, S. M., McPherson, P. S., Mathews, L., Campbell, K. P., and Longo, F. J.,** Cortical localization of a calcium release channel in sea urchin eggs, *J. Cell Biol.,* 116, 1111, 1992.
41. **Brandt, N. R., Caswell, A. H., Brandt, T., Brew, K., and Mellgren, R. L.,** Mapping of the calpain proteolysis products of the junctional foot protein of the skeletal muscle triad junction, *J. Membr.. Biol.,* 127, 35, 1992.
42. **Jayaraman, T., Brillantes, A.-M., Timerman, A. P., Fleischer, S., Erdjument-Bromage, H., Tempst, P., and Marks, A.,** FK506 binding protein associated with the calcium release channel (Ryanodine receptor), *J. Biol. Chem.,* 267, 9474, 1992.
43. **Yang, H. C., Reedy, M. M., Burke, C. L., and Strasburg, G. M.,** Calmodulin interaction with the skeletal muscle sarcoplasmic reticulum calcium channel protein, *Biochemisty,* 33, 518, 1994.
44. **Chen, T. Y. and Yau, K. W.,** Direct modulation of Ca^{2+}-calmodulin of cyclic nucleotide-activated channel of rat olfactory receptor neurons, *Nature,* 368, 545, 1994.
45. **Smith, J. S., Rousseau, E., and Meissner, G.,** Calmodulin modulation of single sarcoplasmic reticulum Ca^{2+}-release channels from cardiac and skeletal muscle, *Circ. Res.,* 64, 352, 1989.
46. **Caterina, M. J. and Devreotes, P. N.,** Molecular insights into eukaryotic chemotaxis, *FASEB J.,* 5, 3078, 1991.
47. **Graeff, R. and Lee, H. C.,** Unpublished results.
48. **Lee, H. C., Graeff, R., Walseth, T. F., Fryxell, K., Branton, W. D., and Lee, H. C.,** Enzymatic synthesis and characterizations of cyclic GDP-ribose. A procedure for distinguishing enzymes with ADP-ribose cyclase activity, *J. Biol. Chem.,* 269, 30260, 1994.

Chapter 4

REGULATION OF RYANODINE RECEPTORS BY CYCLIC ADP-RIBOSE

Antony Galione and Robin Summerhill

TABLE OF CONTENTS

0-8493-8543-1/95/$0.00+$.50

1. INTRODUCTION

The two classes of intracellular Ca^{2+} release channels, inositol 1,4,5-trisphosphate (Ins(1,4,5)P_3) receptors and ryanodine receptors, represent major pathways for liberation of Ca^{2+} from intracellular stores, and the opening of these channels is an important component in the generation of intracellular Ca^{2+} signals. Although ryanodine receptors are key proteins in regulating muscle contraction, their presence in non-muscle cells suggests a more general role in Ca^{2+} homeostasis. Recent evidence indicates that the NAD metabolite, cyclic ADP-ribose, is a potent Ca^{2+} mobilizing agent and acts on a ryanodine-sensitive Ca^{2+} release mechanism. Demonstration of cyclic ADP-ribose (cADPR)-stimulated Ca^{2+} release in several mammalian systems as diverse as neurones, cardiac muscle, and pancreatic cells suggests it may be a general, endogenous modulator of ryanodine-sensitive channels. In this chapter, we review the evidence that cADPR regulates ryanodine receptor-like Ca^{2+} release channels and suggest that it may be a component of a novel Ca^{2+} mobilizing signaling pathway.

2. CYCLIC ADP-RIBOSE: A POTENT ENDOGENOUS Ca^{2+} MOBILIZING AGENT

In view of the many structural and functional similarities between inositol 1,4,5-trisphosphate (Ins(1,4,5)P_3) receptors and ryanodine receptors,[1] it has been suggested that RyRs may be regulated by a small endogenous mediator analagous to Ins(1,4,5)P_3. One candidate that that might fulfill such a role for ryanodine receptors is cyclic ADP-ribose.[2, 3]

cADPR is an endogenous metabolite of β-NAD^+ with potent Ca^{2+} mobilizing activity.[2] It was first discovered and characterized in sea urchin eggs, where addition of NAD^+ to egg homogenates resulted in Ca^{2+} release (Figure 1).[4] A latency period of a number of minutes suggested that NAD^+ was not the active factor per se, but rather was converted into an active metabolite. This was subsequently shown by NMR and mass spectroscopic analysis to be a novel cyclic metabolite, cADPR.[5] cADPR has now been crystallized and previous uncertainties about its structure resolved.[6]

Enzymes involved in the synthesis and degradation of cADPR are found in many invertebrate and mammalian cells.[7,8] ADP-ribosyl cyclases catalyze the synthesis of cADPR from β-NAD^+ and cADPR hydrolases catalyze its

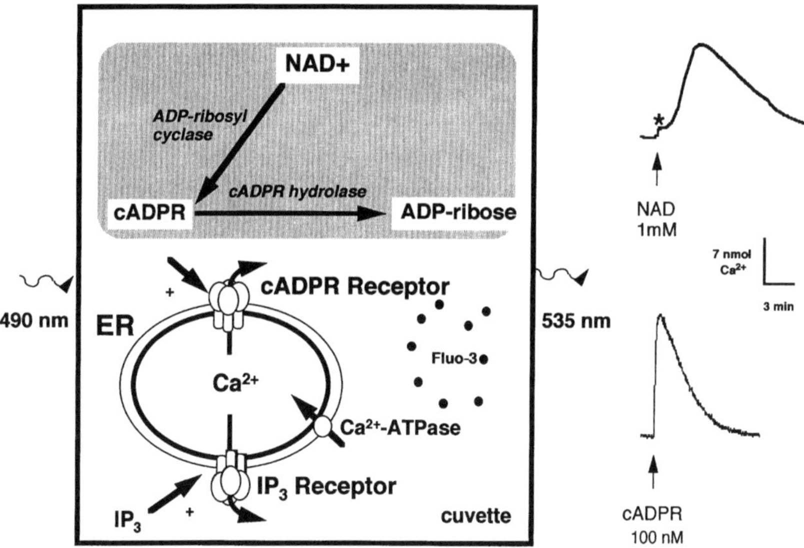

FIGURE 1. Fluorimetric Ca^{2+} release from sea urchin egg homogenates/microsomes. Fluorimetry of egg homogenates using fluorescent Ca^{2+} indicators have allowed a detailed analysis of the metabolism and mechanism of cADPR-induced Ca^{2+} release.[20,23,53,57] Egg homogenates/microsomes are incubated with ATP to allow loading of Ca^{2+} stores via Ca^{2+}-ATPases. Addition of Ca^{2+} mobilizing agents elicits a translocation of Ca^{2+} from stores into the extravesicular medium which is detected by changes in the fluorescence of indicators (depicted as the trace in the lower part of the figure) added to the extravesicular medium in the cuvette. In egg homogenates the metabolic machinery for generating cADPR is still present. NAD-induced Ca^{2+} release occurs after a latency of a minute or so, since it is converted to cADPR by ADP-ribosyl cyclases present in homogenates. cADPR hydrolase is also present which degrades cADPR to its inactive metabolite, ADP-ribose. However, cADPR causes a prompt Ca^{2+} release. Release of Ca^{2+} is often calibrated by addition of known quantities of Ca^{2+} to the cuvette.

degradation to ADP-ribose, which itself has no Ca^{2+} mobilizing activity.[9] ADP-ribosyl cyclase from *Aplysia* ovotestis has been widely studied[8] and has proved particularly useful for the synthesis of cADPR and analogues from NAD^+ and its derivatives.[10] The 29 kDa protein has been immunolocalized to vesicles in the *Aplysia* ovotestis and thus may be considered to be an ectoenzyme.[11]

There is no detectable cADPR-hydrolase activity in *Aplysia* ovotestis. However, in other systems ADP-ribosyl cyclase and cADPR-hydrolase activities appear to be colocalized. It has been reported that in canine spleen both cyclase and hydrolase activities may reside on a single bifunctional 39 kDa polypeptide.[12] The lymphocyte surface antigen, CD38, shows sequence homology with *Aplysia* cyclase[13] and has been shown to possess both cyclase and hydrolase activities in its extracellular domain.[14,15] Antibodies raised against human CD38 induce cell proliferation and maturation depending on the developmental stage of lymphocytes, but it is not known if either the cyclase or hydrolase activities are required for these effects.[16] A homologue of CD38 has been characterized recently from insulinoma cells.[17,18]

These bifunctional enzymes bear a striking relationship with a common class of enzymes known as NAD^+ glycohydrolases or NADases, whose function remains elusive. Such enzymes catalyze the conversion of NAD^+ to ADP-ribose, equivalent to the combined effect of ADP-ribosyl cyclase and cADPR-hydrolase. Indeed, there is an NAD^+ glycohydrolase on the surface of red blood cells which produces cADPR as an intermediate.[19] The significance of the extracellular location of the catalytic domains of these enzymes remains to be determined.[3,15]

3. CYCLIC ADP-RIBOSE-INDUCED Ca^{2+} RELEASE

In a number of cells the pharmacology of cADPR-induced Ca^{2+} release bears striking similarities with that of ryanodine-sensitive release. These similarities have been examined most extensively in sea urchin eggs, but have now been noted in a number of mammalian cell preparations.

3.1. CYCLIC ADP-RIBOSE ACTS ON INS(1,4,5)P_3-INSENSITIVE Ca^{2+} RELEASE CHANNELS

The initial characterization of the Ca^{2+}-mobilizing effects of cADPR in sea urchin egg microsomes demonstrated that it was operating on a Ca^{2+} release mechanism distinct from that activated by Ins(1,4,5)P_3.[4,20] This conclusion was based on two types of studies: desensitization and the use of channel antagonists. Firstly, after Ca^{2+} release induced by cADPR, microsomes were desensitized to subsequent release by an additional dose of cADPR, although they remained sensitive to release by Ins(1,4,5)P_3. Conversely, Ins(1,4,5)P_3-induced Ca^{2+} release appears to desensitize microsomes to further release by Ins(1,4,5)P_3 but not to a subsequent application of cADPR. Secondly, the competitive Ins(1,4,5)P_3 receptor antagonist, heparin, inhibits microsomal release by Ins(1,4,5)P_3, but it has no effect on cADPR-induced release. In contrast, analogues of cADPR, particularly those which have substitutions at the 8-position on the adenine ring, can act as competitive antagonists of cADPR-induced Ca^{2+} release while having no effect on Ins(1,4,5)P_3-induced Ca^{2+} release.[10]

3.2. LOCATION OF CYCLIC ADP-RIBOSE-SENSITIVE STORES

The location of cADPR-sensitive stores has been examined in most detail in the sea urchin egg. cADPR-induced Ca^{2+} release from sea urchin egg homogenates is not affected by mitochondrial inhibitors, indicating that they are situated on non-mitochondrial compartments.[4,20] Fractionation studies of egg homogenates by centrifugation on Percoll gradients revealed that cADPR sensivity as assayed by Ca^{2+} release coincided with microsomal rather than mitochondrial fractions.[20] Furthermore, the cADPR sensitivity paralleled the distribution of microsomal enzyme markers such as glucose-6-phosphatase.

It has not been possible to separate cADPR- and Ins(1,4,5)P_3-sensitive Ca^{2+} pools on Percoll gradients and several lines of experimental evidence suggest that both molecules may be colocalized on microsomal vesicles. Firstly, cADPR and Ins(1,4,5)P_3-induced Ca^{2+} release is nonadditive, suggesting a large overlap between Ins(1,4,5)P_3-sensitive and cADPR-sensitive Ca^{2+} pools in microsomal preparations. Furthermore, confocal microscopy employing eggs loaded with fluorescent Ca^{2+} indicators demonstrates that injection of either cADPR or Ins(1,4,5)P_3 into the egg can induce similar Ca^{2+} waves with no distinct spatial differences.[21]

Although apparently colocalized, Ins(1,4,5)P_3-induced release and cADPR-induced release do represent distinct Ca^{2+} mobilizing pathways. Both release mechanisms show the phenomenon of desensitization; that is, after inducing Ca^{2+} release with a maximal dose of cADPR or Ins(1,4,5)P_3, further Ca^{2+} cannot be released by stimulating with the same agent. However, after maximal Ca^{2+} release by cADPR, egg homogenates are still sensitive to Ins(1,4,5)P_3 and vice versa. This apparent desensitization is not due to depletion of Ca^{2+} pools since it occurs after the released Ca^{2+} has been resequestered. However, it does appear to be dependent on the continued presence of the agonist, since centrifugation of the microsomes and resuspension into fresh medium restores their sensitivity to Ca^{2+} mobilizing agents.[22]

3.3. CYCLIC ADP-RIBOSE AS AN ACTIVATOR OF RYANODINE RECEPTORS

Studies from sea urchin eggs and other cells with cADPR-sensitive Ca^{2+} release mechanisms suggest that cADPR releases Ca^{2+} from intracellular stores by activating ryanodine-sensitive channels (Figure 2). This has raised the provocative hypothesis that in certain cells cADPR is an endogenous modulator of ryanodine receptors.[2,23-26]

The presence of ryanodine receptors in sea urchin eggs was first indicated by functional assays showing ryanodine-induced and caffeine-induced Ca^{2+} release from sea urchin egg microsomes.[23] This was later confirmed by Western blot analysis in which a 380 kDa sea urchin protein cross-reacted with antibodies raised against the mammalian skeletal muscle RyR1.[27]

3.3.1. Cross-Desensitization Studies

Cross-desensitization studies with known ryanodine receptor agonists such as ryanodine and caffeine indicated that cADPR releases Ca^{2+} via a ryanodine-sensitive CICR mechanism.[23] After release of Ca^{2+} by caffeine or ryanodine, the stores were insensitive to cADPR, while Ins(1,4,5)P_3 was still able to elicit a maximal Ca^{2+} release. Conversely, after the Ca^{2+} pools had been discharged by cADPR they were insensitive to further release by cADPR, caffeine, or ryanodine, but fully sensitive to Ins(1,4,5)P_3. These data, together with the ability of the ryanodine receptor blockers, ruthenium red and procaine, to

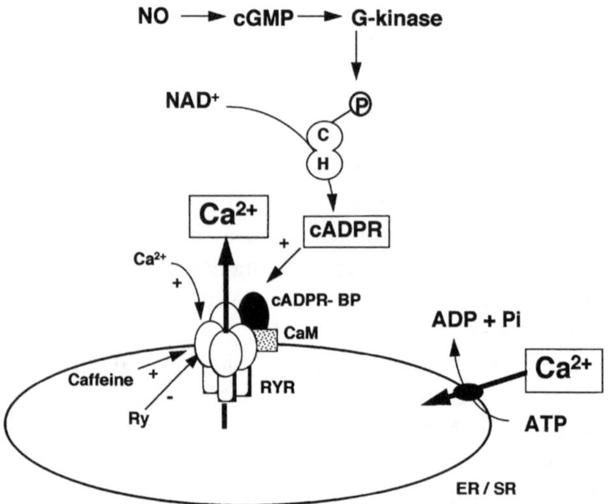

FIGURE 2. Mechanism and control of cADPR-induced Ca^{2+} release. Guanosine 3′,5′-monophosphate (cGMP) can mobilize Ca^{2+} by stimulating cADPR production,[53] as can agents which elevate cGMP in cells such as nitric oxide. A mechanism involving cGMP-dependent protein kinase appears to be involved, which may phosphorylate the bifunctional enzyme which catalyzes both the synthesis and degradation of cADPR leading to an increase in synthesis.[70] Release of Ca^{2+} by cADPR is a complex process and at present poorly understood. Ca^{2+} release by cADPR ultimately appears to be via a ryanodine-like receptor Ca^{2+} release channel,[23] but requires calmodulin,[57] and may indirectly involve the interaction of separate cADPR binding proteins[55] with the ryanodine receptor.

inhibit cADPR but not Ins(1,4,5)P_3-induced Ca^{2+} release, provided the first evidence that cADPR might be a regulator of ryanodine-sensitive Ca^{2+} release channels.[23]

3.3.2. Interactions of cADPR and Ryanodine Receptor Agents

3.3.2.1. Caffeine

Caffeine releases Ca^{2+} from sea urchin egg stores via a ryanodine-sensitive mechanism. The effect is blocked, as in muscle, by ruthenium red or procaine.[23] At millimolar concentrations in the bathing media it can produce a Ca^{2+} transient in the intact egg,[28] although this is rarely sufficient to activate the egg. Caffeine-sensitive Ca^{2+} pools occur not only in the cortex but also deeper into the cytoplasm, and may participate in Ca^{2+} signaling at fertilization and may also regulate later mitotic events.[29]

Interactions between cADPR and caffeine/ryanodine-sensitive Ca^{2+} release mechanisms have been explored in most detail in sea urchin egg homogenates. Caffeine at submaximal concentrations sensitizes homogenates to cADPR-induced Ca^{2+} release.[25] Subthreshold concentrations of 1 to 2 m*M* caffeine can reduce the EC_{50} for cADPR-induced Ca^{2+} release by approximately tenfold. Conversely, low concentrations of cADPR also sensitize Ca^{2+} release mechanisms to activation by caffeine. No such interactions were observed between caffeine and Ins(1,4,5)P_3-induced Ca^{2+} release, consistent with cADPR operating via a ryanodine receptor-like channel rather than an Ins(1,4,5)P_3 receptor.

In a number of mammalian systems, caffeine inhibits Ins(1,4,5)P_3-induced Ca^{2+} release.[30] This is not the case in sea urchin eggs, and may reflect the observations that the sea urchin egg Ins(1,4,5)P_3 receptor is an evolutionary distinct isoform. Indeed, a recent characterization of the sea urchin egg receptor shows that it is larger than mammalian isoforms and interacts differentially with specific mammalian antibodies.[31] In addition, it has extreme salt sensitivity with regard to Ins(1,4,5)P_3 binding.

3.3.2.2. Ryanodine

High concentrations of ryanodine are required to induce Ca^{2+} release either in the intact sea urchin egg or from permeabilized cells, egg cortices, or egg homogenates. Typically, between 100 and 600 μ*M* ryanodine is required, and release is slow compared with that elicited by Ins(1,4,5)P_3 or cADPR. Generally, in intact eggs, ryanodine has to be injected into cells to produce a Ca^{2+} transient, which may[32] or may not[28] result in the exocytotic cortical reaction. However, egg activation has been achieved by bathing eggs in millimolar ryanodine.[33] In both egg cortices and digitonin-permeabilized eggs, ryanodine elicits release of $^{45}Ca^{2+}$ and inhibits its uptake,[33] and can stimulate degranulation of exocytotic vesicles in cortical lawns.[27]

Both the rate and magnitude of Ca^{2+} release induced by ryanodine can be enhanced by low concentrations of caffeine, indicating that the ryanodine receptor in sea urchin eggs is caffeine sensitive.[23] This has also been shown in intact eggs.[28] In addition, subthreshold concentrations of cADPR (10 to 20 n*M*) dramatically sensitize the Ca^{2+} release mechanism to ryanodine, so that lower concentrations of ryanodine (50 μ*M*) which have little effect on their own, can now induce large and quite rapid Ca^{2+} releases.[34] These experiments strongly suggest a close interaction between cADPR and ryanodine receptors.

The effects of ryanodine on sea urchin preparations are generally blocked by ruthenium red and procaine but not by heparin,[28,33] although one report demonstrates that heparin partially blocks the effects of ryanodine[32] while another shows that it actually enhances ryanodine-induced Ca^{2+} release in intact eggs.[28] Enhancement of ryanodine-induced Ca^{2+} release by heparin has been noted, too, in pancreatic acinar cells[35] and heparin has been shown to increase the open probability of ryanodine incorporated into lipid bilayers.[30] It has been suggested that the stimulatory effects of heparin in these instances are due to its high negative charge increasing the local concentration of Ca^{2+} in the vicinity of the channels.

3.3.2.3. Ryanodine Receptor Antagonists

In sea urchin microsomes, where the pharmacology of cADPR has been most extensively studied, classical inhibitors of CICR block cADPR-induced Ca^{2+} release. Ruthenium red and procaine block both caffeine- and cADPR-induced Ca^{2+} release.[23] Interestingly, the cADPR analogue, 8-amino-cADPR blocks, cADPR-induced but not caffeine- or ryanodine-induced Ca^{2+} release.[10] The Ins(1,4,5)P_3 receptor antagonist, heparin, was without effect on either caffeine- or cADPR-induced release.

In many other preparations where cADPR mobilizes Ca^{2+}, ryanodine or ruthenium red blocks cADPR-induced release (Table 1). In pancreatic acinar cells, heparin, which blocks Ins(1,4,5)P_3 -induced release also blocks caffeine and cADPR-induced Ca^{2+}-dependent oscillations in membrane currents.[35] This has been interpreted to suggest that both Ins(1,4,5)P_3 and cADPR are required to generate Ca^{2+} spiking by agonists.

3.3.2.4. Divalent Cations

In sea urchin egg homogenates, high concentrations of Ca^{2+} or Sr^{2+} in excess of 50 m*M* can themselves induce release of sequestered Ca^{2+}.[25] However, threshold concentrations of cADPR sensitize this Ca^{2+}-induced Ca^{2+}release (CICR) mechanism so that release can be induced at 10- to 20-fold less concentration of Ca^{2+} or Sr^{2+}. Similar interactions have been demonstrated with the Ins(1,4,5)P_3 receptor, where Ca^{2+} and Ins(1,4,5)P_3 can act as coagonists.[36-39] Such data indicate that both Ins(1,4,5)P_3 and cADPR/ryanodine-sensitive release mechanisms can support CICR and, indeed, in sea urchin eggs both pathways can independently propagate Ca^{2+} waves at fertilization.[40,41]

Another mode of CICR exists and has been studied in cardiac SR vesicles. Here the loading of stores by Ca^{2+}-ATPases can lead to spontaneous Ca^{2+} release, which is manifested only after a critical filling of the stores has been achieved.[42] The mechanism for this phenomenon is uncertain; one hypothesis is that there is a luminal trigger for the opening of Ca^{2+} release channels which is regulated by the degree of Ca^{2+} loading of the store. It has been proposed as a mechanism for the pathological effects of Ca^{2+} overload in cardiac muscle, whereby Ca^{2+} waves are produced and arrhythmigenic activity is observed.[43] Spontaneous Ca^{2+} release has been shown to occur in non-muscle cells, too, via both Ins(1,4,5)P_3-sensitive[44] and ryanodine-sensitive mechanisms.[45] In sea urchin egg microsomes, spontaneous Ca^{2+} release can occur under conditions of high Ca^{2+} loading of the vesicles. Such release is blocked only when heparin and ruthenium red are added together, indicating that both Ins(1,4,5)P_3R and ryanodine receptors can support this phenomenon. In addition, the degree of Ca^{2+} loading affects the sensitivity of Ca^{2+} release channels to activation by Ca^{2+} mobilizing agents. Loading of stores with sequential Ca^{2+} additions in the presence of ATP can dramatically shift the cADPR dose-response curve to the left with a reduction in EC_{50} of some 20- to 50-fold.[40] A similar sensitization by Ca^{2+} loading was seen for Ins(1,4,5)P_3-induced Ca^{2+} release.

3.3.2.5. Sulfhydryl Reagents

Oxidizing sulfydryl reagents, such as thimerosal or oxidized glutathione, can induce Ca^{2+} release from SR[46] or ER vesicles[44] as well as eliciting Ca^{2+}

TABLE 1
Pharmacology of cADPR-Induced Ca^{2+} Release

Preparation	EC_{50}* or dose range	Heparin	RuRed	Ry	cADPR desens.	[^{3}H]-RY binding	Ca potent	ADPR/heat inactivation
Microsomes	17 n*M**	0	—	—	Yes	+	+	0
Egg		0	—	+	Yes	ND	ND	0
Permeabilized cells	100 n*M*	0	ND	ND	Yes	ND	ND	0
Microsomes	1 μ*M*	0	ND	—	Yes	ND	ND	ND
Acinar cell	100 n*M*	—	—	—	ND	ND	ND	ND
Cell	10 n*M*	ND	0	ND	ND	ND	+	0
Cell	1 μ*M*	ND	ND	—	ND	ND	+	ND
SR	1 μ*M*	ND	ND	—	ND	+	0	ND
RyR2	1 μ*M*	ND	ND	—	ND	ND	+	+
Cell	1 μ*M*	ND	ND	—	ND	ND	ND	0
SR/RYR1	1 μ*M*	ND	0	0	No	—	ND	0
Microsomes	1 μ*M*	ND	0	ND	Yes	ND	ND	0
Microsomes/ vacuole	24 n*M**	ND	—	—	ND	ND	ND	0

Note: The data demonstrates the similarity in pharmacology between ryanodine receptor-mediated Ca^{2+} release and that activated by cADPR. The effective dose-range or EC_{50} for the effects are shown. Several criteria are used to demonstrate cADPR-induced Ca^{2+} release. It is generally ruthenium red-sensitive but not blocked by heparin. Ryanodine inhibits release and [^{3}H]-ryanodine binding may be modulated by cADPR. Often cADPR-induced release becomes refractory to further release. Heat treatment of cADPR yields ADP-ribose, which does not cause release of Ca^{2+}. In many preparations cADPR modulates CICR, and thus Ca^{2+} potentiates cADPR-induced Ca^{2+} release. The table demonstrates whether these criteria are satisfied in different systems. Concordance or deviations from these are discussed in the text. Key: 0 no effect; – inhibits; + potentiates; ND not determined. In one report,[58] cADPR was shown to be ineffective at modulating RYR1 in lipid bilayers. However, cADPR at high micromolar concentrations released calcium from skeletal muscle SR vesicles. Such release did not 'desensitize' and was insensitive to ryanodine receptor blockers such as ruthenium red.

spiking or Ca^{2+} waves in cells as diverse as mammalian eggs[47] or liver cells.[48] In permeabilized liver cells, the effects of such sulphydryl reagents are similar to those of Ca^{2+} loading, in that they evoke a slow but accelerating release of stored Ca^{2+}, and that they sensitize stores to release by Ins(1,4,5)P_3. Both effects, in the case of hepatocytes, can be blocked by heparin,[44] suggesting that the mechanism for such actions may occur at the level of the Ins(1,4,5)P_3 receptor, while the effects in SR indicate that ryanodine receptors may be involved.[46] In sea urchin egg homogenates, thimerosal can sensitize the cADPR-induced Ca^{2+} release mechanism; at concentrations of thimerosal that cause no release themselves (20 μ*M*), subthreshold doses of cADPR (10 n*M*) can now cause substantial Ca^{2+} release. In intact sea urchin eggs, a twofold sensitization of cADPR-induced Ca^{2+} release was seen in thimerosal-treated eggs.[49] Similar results are seen for sensitization of Ins(1,4,5)P_3-induced Ca^{2+} release in this system,[40,50] further exemplifying the similarities between the two Ca^{2+} release pathways.

Thimerosal-induced Ca^{2+} release is only blocked when ruthenium red and heparin are added together, further reinforcing the idea that sulfydryl reagents can modify both ryanodine and Ins(1,4,5)P_3 receptors.[40] However, in another report, procaine and heparin together were not sufficient to block thimerosal-induced Ca^{2+} release.[50] In this latter report, however, high concentrations of thimerosal were used, which may inhibit Ca^{2+} sequestration by the microsomes, since the time course of release was similar to that induced by thapsigargin, a potent microsomal Ca^{2+}-ATPase inhibitor.

4. THE CYCLIC ADP-RIBOSE RECEPTOR

Pharmacological evidence from sea urchin studies suggests that cADPR acts via a ryanodine-sensitive Ca^{2+} release mechanism. Perhaps the simplest model for this interaction is that cADPR binds directly to a ryanodine receptor, acting as an endogenous activator analagous to Ins(1,4,5)P_3. Thus, the ryanodine receptor and cADPR receptor are synonymous. To confirm or refute this hypothesis, we must understand the nature of the sea urchin ryanodine receptor.

The presence of a ryanodine receptor-like protein in sea urchin eggs has been biochemically demonstrated, as stated previously, by Western blot analysis.[27] However, the characteristics of this protein remain to be elucidated, and it is not known if it correlates with any of the known mammalian isoforms.[51, 52] The apparent molecular weight of this protein (380 kDa) is considerably smaller than the individual subunit size of the three main RyRs found in mammalian cells. In addition, its localization to the egg cortex contradicts evidence from confocal fluorescent microscopy which indicates that ryanodine receptors can mediate the propagation of Ca^{2+} waves deep into the egg cytoplasm.[29,53] Whether this protein is modulated by cADPR is unknown.

Two strategies have been employed to characterize the cADPR receptor in sea urchin egg microsomes. The first involves measuring binding of radiolabeled or photoaffinity analogues of cADPR to microsomal proteins.[54,55] The second is to examine the interactions between cADPR and ryanodine receptors by monitoring changes in ryanodine binding and Ca^{2+} release characteristics.[26]

4.1. BINDING STUDIES OF [^{32}P]-CYCLIC ADP-RIBOSE

In all cellular systems that have been shown to be reponsive to cADPR, the metabolite is effective at nanomolar concentrations, indicating a receptor system is involved. Investigation of a possible receptor interaction has been made possible through the availability of high specific activity radiolabeled cADPR produced by ADP-ribosyl cyclase-catalyzed conversion of [^{32}P]-NAD^+.[8] Saturatable and specific binding of cAPDR to sea urchin egg microsomes purified by Percoll density centrifugation has been demonstrated.[54] Scatchard analysis showed that the receptor density (B_{max}) is about 25 fmol/mg protein and the dissociation constant (K_D) is about 17 n*M*, which is very similar to the half-maximal concentration for Ca^{2+} release. Ins(1,4,5)P_3 was 1000 times less effective than cADPR in competing for the binding site, showing that the receptor for cADPR is distinct from the Ins(1,4,5)P_3 receptor. Furthermore, heparin has very little effect on cADPR binding.[54]

4.2. CYCLIC ADP-RIBOSE ANTAGONISTS

In an effort to understand the molecular mechanism of how cADPR activates Ca^{2+} release and to determine the physiological role of the metabolite, a series of analogues of cADPR were synthesized and tested for their ability to induce Ca^{2+} release in sea urchin egg microsomes.[10] The 8-position of the adenine ring of cADPR was found to be of crucial importance. Any modification at this position eliminates the Ca^{2+} release activity of the analogue. However, the modified analogues can still bind to the microsomal receptor site and act as competitive antagonists of cADPR-induced Ca^{2+} release. Among the analogues synthesized, 8-amino-cADPR is the most potent antagonist, reducing the rate of Ca^{2+} release induced by 135 n*M* cADPR by 50% when present at 9 n*M*. Although 8-amino-cADPR was very effective in blocking Ca^{2+} release induced by cADPR, it had no effect on the Ins(1,4,5)P_3-sensitive release. Heparin, on the other hand, blocked the Ins(1,4,5)P_3-induced release but not that of cADPR. These results show that there are definitely two different Ca^{2+} release systems in sea urchin eggs — one is sensitive cADPR and the other to Ins(1,4,5)P_3.

8-Amino-cADPR does not block caffeine or ryanodine-induced Ca^{2+} release from egg microsomes. This suggests that cADPR, ryanodine, and caffeine act at different binding sites on the ryanodine receptor or that cADPR binds to an accessory protein which then interacts with the ryanodine receptor. If the cADPR binding site is not the ryanodine receptor itself, it must nevertheless be tightly associated with the ryanodine receptor since cADPR is equally effective at releasing Ca^{2+} from sea urchin egg homogenates and from purified microsomes.[20] The 8-substituted analogues of cADPR are certain to become important tools in unraveling the physiological roles of cADPR.

4.3. PHOTOAFFINITY LABELING

Although specific microsomal binding sites for cADPR have been demonstrated, the nature of these sites remains unclear. To further characterize these sites, a photoactive analogue of cADPR, 8-azido-cADPR, has been synthesized.[55] Similar to 8-amino-cADPR, the azido derivative is also an antagonist and can effectively compete for the cADPR binding sites. High specific activity ^{32}P-labeled 8-azido-cADPR has also been synthesized. Binding of the probe to egg microsomes can be specifically blocked by nanomolar concentrations of cADPR, indicating that both are binding to the same sites.

Photoactivation of [^{32}P]-8-azido-cADPR resulted in specific labeling of two sea urchin microsomal proteins with molecular weights of 140 and 100 kDa.[55] The presence of nanomolar concentrations of cADPR completely blocked the labeling of these proteins, while other adenine nucleotides such as ATP, ADP, AMP, cAMP, and ADP-ribose had no effect even at micromolar concentrations. Interestingly, caffeine (5 to 10 m*M*), the classical agonist of CICR, selectively and preferentially reduced the photoaffinity labeling of the 100 kDa protein as compared to the 140 kDa protein, suggesting that the former may be the target protein of caffeine.

The molecular weights of the two labeled proteins are four to five times smaller than expected for known subunits of ryanodine receptors.[51,52] It is possible that the 140 kDa and 100 kDa proteins may represent variant forms of the ryanodine receptor. Indeed, a recent report has suggested that a brain-specific transcript of the skeletal muscle ryanodine receptor gene encodes for a protein of about 70 kDa.[56] Alternatively, the140 kDa and 100 kDa proteins may act as accessory proteins, mediating the interaction of cADPR with the ryanodine receptor. In any case, the success of photoaffinity labeling the cADPR binding proteins should set the stage for their eventual purification and identification. It may turn out that cADPR-induced Ca^{2+} release is a multistep process involving several proteins acting in concert, a process much more complicated than the action of Ins(1,4,5)P_3.[57]

4.4. RYR2 AS THE CYCLIC ADP-RIBOSE RECEPTOR

Studies of the pharmacology of cADPR-induced Ca^{2+} release have been extended to a number of mammalian systems. In most cases where cADPR-induced Ca^{2+} release can be demonstrated, ryanodine blocks this release, whereas antagonists of Ins(1,4,5)P_3 receptors have no inhibitory effects.

Evidence suggests that cADPR interacts with the mammalian cardiac isoform (RyR2) but not the skeletal isoform (RyR1). cADPR like caffeine enhances [^{3}H]-ryanodine binding to cardiac but not to skeletal muscle SR membranes.[26]

The hypothesis that cADPR modulates RyR2 channel opening has been directly tested using ryanodine receptors incorporated into lipid bilayers. In one study, the finding that cADPR enhanced Ca^{2+} release from cardiac SR vesicles was extended to single-channel studies in which SR vesicles were incorporated into artificial lipid bilayers.[26] Activation of RyR2 channels in this system by maximal concentrations of cADPR (1 μ*M*) was most pronounced at low Ca^{2+} concentrations (10 to 100 n*M*) although divalent cations clearly potentiate submaximal cADPR-induced Ca^{2+} release from sea urchin egg microsomes[25] (see Section 3.3.2.4).These studies were performed with SR vesicles and not purified RyR2 receptors, thus one cannot discount the possibility that cADPR binds first to an accessory protein forming a complex which then interacts with the ryanodine receptor. The observed effect on the receptor's affinity for ryanodine may be due to a conformational change resulting from interactions between the receptor and cADPR directly or interactions between an accessory complex and the receptor.

The effects of cADPR on the skeletal muscle RyR1 receptor have proved negative[23,26,58] with the exception that cADPR may reduce [^{3}H]-ryanodine binding to this isoform.[26]

The effects of cADPR on the RyR3 isoform have yet to be determined. However, since this isoform is relatively insensitive to caffeine,[59] it seems unlikely that the RyR3 isoform is a major target for cADPR in view of the strong interactions between caffeine and cADPR.

4.5. REQUIREMENT FOR CYTOSOLIC FACTORS

Characterization of the ryanodine receptor/Ca^{2+} release channel complex has demonstrated that additional accessory proteins are required for the normal function of the channel. For example, small proteins which bind the immunosuppressant drug, FK-506, known as FK-506 binding proteins, are required for the proper, coordinated opening of channels from skeletal muscle when incoporated into lipid bilayers.[60] Additional protein factors are also required to confer cADPR sensitivity on purified microsomes. In the first study to demonstrate a Ca^{2+} mobilizing effect of an NAD^+ metabolite on sea urchin egg microsomes, it was noted that in addition to an enzyme activity (presumably ADP-ribosyl cyclase), additional supernatant factors were required to reconstitute NAD^+-induced Ca^{2+} release.[4] Such factors could be neccessary for the activity of ADP-ribosyl cyclase or for the action of cADPR. If the latter case is correct, then possible candidates include cADPR binding proteins distinct from the Ca^{2+} release channel[55] or modulatory proteins required in addition to cADPR. Indeed, a supernatant factor, recently identified as calmodulin, has been shown to be essential for cADPR-induced Ca^{2+} release from purified sea urchin egg microsomes.[57] This is a surprising finding, since calmodulin has been shown to reduce the RyR1 and RyR2 opening probability and Ca^{2+} release from skeletal and cardiac SR.[51]

Since cADPR has generally been shown to produce more robust effects in intact cells, it is possible that supernatant factors such as calmodulin may be lost or depleted in permeabilized cell systems and microsomal preparations. This could be one explanation for the difficulties some authors have experienced when trying to demonstrate cADPR effects in certain broken mammalian cell preparations.[61]

5. CYCLIC ADP-RIBOSE AS A REGULATOR OF RYANODINE RECEPTORS IN MAMMALIAN CELLS

The finding that cADPR itself together with ADP-ribosyl cyclases and hydrolases are present in a wide range of mammalian systems[3] prompted the investigation of cADPR effects on Ca^{2+} mobilization in mammalian cells. In all preparations which display cADPR-induced Ca^{2+} release, caffeine or ryanodine-sensitive Ca^{2+} release mechanisms are present and this has proved a useful indicator of candidate systems for study.

cADPR has been shown to mobilize Ca^{2+} from permeabilized pituitary cells,[62] patch-clamped dorsal root ganglion cells,[63] pancreatic β cells,[64] brain microsomes,[26,64,65] and heart SR vesicles,[26] indicating a widespread role in controlling Ca^{2+} signals. In all these cell types, release occurs via an Ins(1,4,5)P_3-insensitive mechanism. Furthermore, cADPR-induced Ca^{2+} release from brain microsomes[26,64,65] was blocked by ryanodine, as was release from pancreatic β cell microsomes[64] and cardiac SR vesicles,[26] suggesting that cADPR, as in the sea urchin egg, operates via a ryanodine receptor. At nanomolar concentrations, cADPR, but not Ins(1,4,5)P_3, mimics caffeine in inducing Ca^{2+} oscillations in rat dorsal root ganglion cells, as monitored by a Ca^{2+}-dependent ion currents known to reflect intracellular Ca^{2+} oscillations.[63] This action of cADPR in these cells was potentiated by a Ca^{2+} influx which raised intracellular Ca^{2+} levels, and was abolished by prior depletion of caffeine-sensitive stores, consistent with cADPR modulating CICR via a ryanodine receptor-mediated mechanism. Similar results were observed in bullfrog sympathetic neurones, which also contain a caffeine-sensitive Ca^{2+} pool.[66] A recent report has shown that intracellular perfusion of cells with cADPR induces short-lived Ca^{2+} spikes in the apical region of cells.[35] The frequency of Ca^{2+} spikes increases with increasing pipette concentration of cADPR. Ryanodine abolishes these cADPR-induced spikes and reduces agonist-induced responses, but does not reduce the effects of Ins(1,4,5)P_3 and may even enhance them.[35] A complication in this study is that heparin not only abolishes the effects of Ins(1,4,5)P_3 but also abolishes the effects of caffeine and cADPR, in contrast to its selective inhibition of Ins(1,4,5)P_3-induced Ca^{2+} release in sea urchin eggs. This has been interpreted as evidence that the spatiotemporal complexity in Ca^{2+} signals in exocrine pancreatic cells results from the interplay between Ins(1,4,5)P_3 receptors and ryanodine receptors.

6. CYCLIC ADP-RIBOSE: Ca^{2+}-MOBILIZING SECOND MESSENGER OR MODULATOR OF CICR?

Despite the growing number of reports of the effects of cADPR on Ca^{2+} release in cells, little is known about the control of cADPR synthesis. A major unresolved question is whether cADPR might play an analogous role to Ins (1,4,5)P_3 in Ca^{2+} signaling by functioning as a second messenger whose synthesis is coupled to the acute activation of cell surface receptors, or whether it is an intracellular modulator of Ca^{2+} release channels whose levels are linked to general metabolic processes in the cell. The latter may be the case in bullfrog sympathetic neurones. In these cells the primary stimulus for CICR is the influx of Ca^{2+} across the plasma membrane, the effect of cADPR is to sensitize the CICR mechanism to this stimulus. Although cADPR satisfies several of the criteria for a true second messenger, in no system have they all been demonstrated. The most obvious gap is that there is at present little evidence to suggest that cADPR levels change in stimulated cells. This is mainly due to the lack, at present, of a sensitive, straightforward method of measuring cADPR in cells. However, there is preliminary evidence in two systems — the sea urchin egg and the pancreatic β cell — that cADPR may behave as a second messenger.

6.1. CYCLIC ADP-RIBOSE AS A SECOND MESSENGER

6.1.1. Sea Urchin Egg

Evidence suggests that cADPR has a direct physiological role in the production and propagation of Ca^{2+} waves at fertiliztion in sea urchin eggs. An Ins(1,4,5)P_3-sensitive Ca^{2+} release mechanism is involved in the generation of this fertilization wave. However, studies in eggs injected with ruthenium red and 8-amino-cADPR have implicated a cADPR/ryanodine receptor-mediated mechanism in the process.[40,41] Both Ins(1,4,5)P_3-sensitive and cADPR-sensitive CICR mechanisms must be blocked to completely abolish the fertilization wave and appear to act in a redundant manner.

It has been suggested that the synthesis of cADPR may be regulated by cGMP in the sea urchin egg. Injection of cGMP into eggs causes a large Ca^{2+} transient which is independent of extracellular Ca^{2+}.[67] As with fertilization, but in contrast to Ca^{2+} mobilization by Ins(1,4,5)P_3 or cADPR, there is a latency of several seconds before initiation of the Ca^{2+} transient. The effect of cGMP is blocked by the prior release of Ca^{2+} by cADPR, ruthenium red,[53] and 8-amino-cADPR, but not by heparin. In addition, cGMP is only able to release Ca^{2+} in the presence of exogenous NAD^+.[53] This suggested that cGMP stimulates the conversion of β-NAD^+ to cADPR and, indeed, this has been indicated by separation of radiolabeled NAD^+ and metabolites by thin-layer chromatography.

The cGMP effect is also blocked by protein kinase inhibitors, suggesting that a cGMP-dependent phosphorylation stimulates cADPR production, although the mechanism remains to be clarified. Interestingly, NO and the NO-donor hydroxylamine induce Ca^{2+} release in sea urchin eggs and in egg homogenates (Wilmot, N., Sethi, J., White, A., Walsern, T., Lee, H. C., and Galione, A., unpublished results). This is NAD^+ dependent, is inhibited by 8-amino-cADPR, and the effect is reduced considerably by the presence of inhibitors of cGMP-dependent kinase. Although it remains to be demonstrated directly that ADP ribosyl cyclase is activated by cGMP-dependent phosphorylation, consensus sites for phosphorylation have been identified in the sequence from the lymphocyte antigen CD38,[3] which has both cyclase activity and shares considerable sequence homology with the *Aplysia* cyclase.

Although cADPR is required for ryanodine receptor-mediated Ca^{2+} wave propagation, and though its synthesis can be regulated, it remains to be demonstrated that fertilization leads to increased cADPR levels and, hence, functions as a classical second messenger at fertilization.

6.1.2. Endocrine Pancreas

As described earlier there is evidence that cADPR is active in pancreatic β cells, although this remains controversial.[68,69] Nevertheless, a possible role in mediating stimulus-secretion coupling is indicated by the finding that cytoplasmic extracts from islets which have been treated with high concentrations of glucose contain a factor which releases Ca^{2+} from cerebellar and pancreatic β cell microsomes. This release was shown to be mediated by a cADPR-sensitive mechanism.[64] Furthermore, diabetogenic drugs which deplete cells of the cADPR precursor, NAD^+, and inhibit insulin secretion, also prevent the production of this Ca^{2+}-releasing factor, which has been tentatively assigned as cADPR. This has been taken as evidence for cADPR mediating glucose-induced insulin secretion,[2,64] although biochemical measurements of cADPR in stimulated β cells have yet to be carried out. One possible mechanism for coupling glucose stimulation of β cells with cADPR synthesis has been suggested.[17] The cADPR hydrolase activity of a CD38-like molecule demonstrated in insulinoma cells is inhibited by ATP. An increase in ATP on exposure to glucose may lead to inhibition of this cADPR hydrolase activity and an increase in levels of cADPR.

7. CONCLUSIONS

The widespread occurrence of cADPR and the enzymes involved in its metabolism, as well as early reports of cADPR-induced Ca^{2+} release in several mammalian cells, indicates that cADPR may have an important role in intracellular Ca^{2+} signaling. The molecular identification of the cADPR binding site and its precise relationship to ryanodine-sensitive Ca^{2+} release channels has yet to be determined. It is not known whether the major target of cADPR is any of the previously characterized ryanodine receptors, or whether it modulates a novel, as yet uncharacterized isoform. This may explain in part some of the conflicting reports on whether cADPR has Ca^{2+} mobilizing activity in some cells. What is clear is that the pharmacology of cADPR-induced Ca^{2+} release in several different sytems suggests that it is an activator of ryanodine receptors rather than Ins(1,4,5)P_3 receptors. Whether cADPR functions mainly as a modulator of CICR or whether it can act as a classical second messenger, being synthesized upon activation of cells by extracellular stimuli, remains to be determined. A more complete knowledge of the mechanisms of cADPR-induced Ca^{2+} release and the effects of extracellular agonists on intracellular levels of cADPR should clarify the role of cADPR in cell signaling.

ACKNOWLEDGMENTS

We wish to thank the Medical Research Council, Biotechnology and Biological Sciences Research Council, and Wellcome Trust for support. AG is a Wellcome Fellow and RS is a Christopher Welch Scholar.

REFERENCES

1. **Furuichi, T., Kohda, K., Miyawaki, A., and Mikoshiba, K.,** Intracellular channels, *Curr. Op. Neurobiol.*, 4, 294, 1994.
2. **Galione, A.,** Cyclic ADP-ribose, a new way to control calcium, *Science*, 259, 325, 1993.
3. **Lee, H. C., Galione, A., and Walseth, T. F.,** Cyclic ADP-ribose: metabolism and calcium mobilizing function, in *Vitamins and Hormones,* Litwack, G., Ed., Academic Press, Orlando, FL, 1994, 199.
4. **Clapper, D., Walseth, T., Dargie, P., and Lee, H. C.,** Pyridine nucleotide metabolites stimulate calcium release from sea urchin egg microsomes desensitized to inositol trisphosphate, *J. Biol. Chem.*, 262, 9561, 1987.
5. **Lee, H. C., Walseth, T. F., Bratt, G. T., Hayes, R. N., and Clapper, D. L.,** Structural determination of a cyclic metabolite of NAD^+ with intracellular Ca^{2+}-mobilizing activity, *J. Biol. Chem.*, 264, 1608, 1989.
6. **Lee, H. C., Aarhus, R., and Levitt, D.,** The crystal structure of cyclic ADP-ribose, *Nature Struct. Biol.*, 1, 143, 1994.
7. **Rusinko, N. and Lee, H. C.,** Widespread occurrence in animal tissues of an enzyme catalyzing the conversion of NAD^+ into a cyclic metabolite with intracellular Ca^{2+}-mobilizing activity. *J. Biol. Chem.*, 264, 11725, 1989.
8. **Lee, H. C. and Aarhus, R.,** ADP-ribosyl cyclase: an enzyme that cyclizes NAD^+ into a calcium-mobilizing metabolite, *Cell Regul.*, 2, 203, 1991.
9. **Lee, H. C. and Aarhus, R.,** Widespread distribution of an enzyme that catalyses the hydrolysis of cyclic ADP-ribose, *Biochim. Biophys. Acta*, 1164, 68, 1993.
10. **Walseth, T. F. and Lee, H. C.,** Synthesis and characterization of antagonists of cyclic-ADP-ribose-induced Ca^{2+} release, *Biochim. Biophys. Acta*, 1178, 235, 1993.
11. **Hellmich, M. and Strumwasser, F.,** Purification and characterization of a molluscan egg-specific NADase, a second-messenger enzyme, *Cell Regul.*, 2, 193, 1991.
12. **Kim, H., Jacobson, E. L., and Jacobson, M. K.,** Synthesis and degradation of cyclic ADP-ribose by NAD glycohydrolases, *Science*, 261, 1330, 1993.
13. **States, D., Walseth, T., and Lee, H. C.,** Similarities in amino acid sequences of *Aplysia* ADP-ribosyl cyclase and human lymphocyte antigen CD38, *Trends Biochem. Sci.*, 17, 495, 1992.
14. **Howard, M., Grimaldi, J. C., Bazan, J. F., Santos-Argumedo, L., Parkhouse, R. M. E., Walseth, T. F., and Lee, H. C.,** Lymphocyte antigen CD38 catalyzes the formation and hydrolysis of cyclic ADP-ribose, *Science*, 262, 1056, 1993.
15. **Summerhill, R. J., Jackson, D. G., and Galione, A.,** Human lymphocyte antigen CD38 catalyzes the production of cyclic ADP-ribose, *FEBS Lett.*, 335, 231, 1993.
16. **Malavasi, F., Funaro, A., Roggero, S., Horenstein, A., Calosso, L., and Mehta, K.,** Human CD38: a glycoprotein in search of a function, *Immunol., Today,* 15, 95, 1994.
17. **Takasawa, S., Tohgo, A., Noguchi, N., Koguma, T., Nata, K., Sugimoto, T., Yonekura, H., and Okamoto, H.,** Synthesis and hydrolysis of cyclic ADP-ribose by human leukocyte antigen CD38 and inhibition of hydrolysis by ATP, *J. Biol. Chem.*, 268, 26052, 1993.
18. **Koguma, T., Takasawa, S., Tohgo, A., Karasawa, T., Furuya, Y., Yonekura, H., and Okamoto, H.,** Cloning and characterization of cDNA encoding rat ADP-ribosyl cyclase/cyclic ADP-ribose hydrolase (homologue to human CD38) from islets of Langerhans, *Biochim. Biophys. Acta,* 1223, 160, 1994.
19. **Lee, H. C., Zocchi, E., Guida, L., Franco, L., Benatti, U., and De Flora, A.,** Production and hydrolysis of cyclic ADP-ribose at the outer surface of human erythrocytes, *Biochim. Biophys. Res. Commun.*, 191, 639, 1993.
20. **Dargie, P. J., Agre, M. C., and Lee, H. C.,** Comparison of Ca^{2+} mobilizing activities of cyclic ADP-ribose and inositol trisphosphat, *Cell Regul.*, 1, 279, 1990.

21. **Shen, S. S. and Buck, W. R.,** Sources of calcium in sea urchin eggs during the fertilization response, *Dev. Biol.*, 157, 1993.
22. **Clapper, D. L. and Lee, H. C.,** Inositol trisphosphate induces Ca^{2+}- release from nonmitochondrial stores in sea urchin egg homogenates, *J. Biol. Chem.*, 260, 13947, 1985.
23. **Galione, A., Lee, H. C., and Busa, W. B.,** Ca^{2+}-induced Ca^{2+} release in sea urchin egg homogenates: modulation by cyclic ADP-ribose, *Science*, 253, 1143, 1991.
24. **Galione, A.,** Ca^{2+}-induced Ca^{2+} release and its modulation by cyclic ADP-ribose, *Trends Pharmacol. Sci.*, 13, 304, 1992.
25. **Lee, H. C.,** Potentiation of calcium- and caffeine-induced calcium release by cyclic ADP-ribose, *J. Biol. Chem.*, 268, 293, 1993.
26. **Meszaros, L. G., Bak, J., and Chu, A.,** Cyclic ADP-ribose as an endogenous regulator of the non-skeletal type ryanodine receptor Ca^{2+} channel, *Nature*, 364, 76, 1993.
27. **McPherson, S. M., McPherson, P. S., Mathews, L., Campbell, K. P., and Longo, F. J.,** Cortical localization of a calcium release channel in sea urchin eggs, *J. Cell Biol.*, 116, 1111, 1992.
28. **Buck, W., Rakow, T., and Shen, S.,** Synergistic release of calcium in sea urchin eggs by caffeine and ryanodine, *Exp. Cell. Res.*, 202, 59, 1992.
29. **Harris, P. J.,** Caffeine-induced calcium release in sea urchin eggs and the effect of continuous versus pulsed application on the mitotic apparatus, *Dev. Biol.*, 161, 370, 1994.
30. **Ehrlich, B. E., Kaftan, E., Bezprozvannaya, S., and Bezprozvanny, I.,** The pharmacology of intracellular Ca^{2+} release channels, *Trends Pharmacol. Sci.*, 15, 145, 1994.
31. **Parys, J. B., McPherson, S. M., Mathews, L., Campbell, K. P., and Longo, F. J.,** Presence of inositol 1,4,5-trisphosphate receptor calreticulin and calsequestrin in eggs of sea urchins and *Xenopus laevis, Dev. Biol.*, 161, 466, 1994.
32. **Sardet, C., Gillot, I., Ruscher, A., Payan, P., Girard, J.-P., and de Renzis, G.,** Ryanodine activates sea urchin eggs, *Develop. Growth Differ.*, 34, 37, 1992.
33. **Fujiwara, A., Taguchi, K., and Yasumasu, I.,** Fertilization membrane formation in sea urchin eggs induced by drugs known to cause Ca^{2+} release from isolated sarcoplasmic reticulum, *Develop. Growth Differ.*, 32, 303, 1990.
34. **Buck, W. R., Hoffmann, E. E., Rakow, T. L., and Shen, S. S.,** Synergistic calcium release in the sea urchin egg by ryanodine and cyclic ADP-ribose, *Dev. Biol.*, 163, 1, 1994.
35. **Thorn, P., Gerasimenko, O., and Petersen, O. H.,** Cyclic ADP-ribose regulates ryanodine receptors involved in agonist-evoked cytosolic Ca^{2+} oscillations, *EMBO J.*, 13, 2038, 1994.
36. **Bezprozvanny, I., Watras, J., and Ehrlich, B. E.,** Bell-shaped calcium-response curves of Ins(1,4,5)P3- and calcium-gated channels from endoplasmic reticulum of cerebellum, *Nature*, 351, 751, 1991.
37. **Finch, E. A., Turner, T. J., and Goldin, S. M.,** Calcium as a coagonist of inositol 1,4,5-trisphosphate-induced calcium release, *Science*, 252, 443, 1991.
38. **Iino, M. and Endo, M.,** Calcium-dependent immediate feedback control of inositol 1,4,5 trisphosphate-induced calcium release, *Nature*, 360, 76, 1992.
39. **Berridge, M. J. and Dupont, G.,** Spatial and temporal signaling by calcium, *Curr. Op. Cell Biol.*, 6, 267, 1994.
40. **Galione, A., McDougall, A., Busa, W., Willmott, N., Gillot, I. and Whitaker, M.,** Redundant mechanisms of calcium-induced calcium release underlying calcium waves during fertilization of sea urchin eggs, *Science*, 261, 348, 1993.
41. **Lee, H. C., Aarhus, R., and Walseth, T. F.,** Calcium mobilization by dual receptors during fertilization of sea urchin eggs, *Science*, 261, 352, 1993.
42. **Palade, P., Dettbarn, C., Brunder, D., Stein, P., and Hals, G.,** Pharmacology of calcium release from sarcoplasmic reticulum, *J. Bioen. Biomemb.*, 21, 295, 1989.
43. **Lipp, P. and Niggli, E.,** Modulation of Ca^{2+} release in cultured neonatal rat cardiac myocytes. Insight from subcellular release patterns revealed by confocal microscopy, *Circ. Res.*, 74, 979, 1994.

44. **Missiaen, L., Taylor, C., and Berridge, M.,** Spontaneous calcium release from inositol trisphosphate-sensitive calcium stores, *Nature*, 352, 241, 1991.
45. **Cheek, T. R., Barry, V. A., Berridge, M. J., and Missiaen, L.,** Bovine adrenal chromaffin cells contain an inositol 1,4,5 trisphosphate-insensitive but caffeine-sensitive Ca^{2+} store that can be regulated by intraluminal free Ca^{2+}, *Biochem. J.*, 275, 697, 1991.
46. **Zaidi, N. F., Lagenauer, C. F., Hilkert, R. J., Xiong, H., Abramson, J. J., and Salama, G.,** Disulfide-linkage of biotin identifies a 106 kDa Ca^{2+} release channel in sarcoplasmic reticulum, *J. Biol. Chem.*, 264, 21737, 1989.
47. **Swann, K.,** Thimerosal causes calcium oscillations and sensitizes calcium-induced calcium release in unfertilized hamster eggs, *FEBS Lett.*, 278, 175, 1991.
48. **Rooney, T. A., Renard, D. C., Sass, E. J., and Thomas, A. P.,** Oscillatory cytosolic calcium waves independent of stimulated inositol 1,4,5-trisphosphate formation in hepatocytes, *J. Biol. Chem.*, 266, 12272, 1991.
49. **McDougall, A., Gillot, I., and Whitaker, M. J.,** Thimerosal reveals calcium-induced calcium release in unfertilized sea urchin eggs, *Zygote*, 1, 35, 1993.
50. **Tanaka, Y. and Tashjian, A. H.,** Thimerosal potentiates Ca^{2+} release mediated by both the inositol 1,4,5-trisphosphate and the ryanodine receptors in sea urchin eggs — implications for mechanistic studies on Ca^{2+} signaling, *J. Biol. Chem.*, 269, 11247, 1994.
51. **Meissner, G.,** Ryanodine receptor Ca^{2+} release channels and their regulation by endogenous effectors, *Annu. Rev. Physiol.*, 56, 485, 1994.
52. **Coronado, R., Morrissette, J., Sukhareva, M., and Vaughan, D. M.,** Structure and function of ryanodine receptors, *Am. J. Physiol.*, 266, C1485, 1994.
53. **Galione, A., White, A., Willmott, N., Turner, M., Potter, B. V., and Watson, S. P.,** cGMP mobilizes intracellular Ca^{2+} in sea urchin eggs by stimulating cyclic ADP-ribose synthesis, *Nature*, 365, 456, 1993.
54. **Lee, H. C.,** Specific binding of cyclic ADP-ribose to calcium-storing microsomes from sea urchin eggs, *J. Biol. Chem.*, 266, 2276, 1991.
55. **Walseth, T. F., Aarhus, R., Kerr, J. A., and Lee, H. C.,** Identification of cyclic ADP-ribose-binding proteins by photoaffinity labeling, *J. Biol. Chem*,. 268, 26686, 1993.
56. **Takeshima, H., Nishimura, S., Nishi, M., Ikeda, M., and Sugimoto, T.,** A brain-specific transcript from the 3′-terminal region of the skeletal muscle ryanodine receptor gene, *FEBS Lett.*, 322, 105, 1993.
57. **Lee, H. C., Aarhus, R., Graeff, R., Gurnack, M. E., and Walseth, T. F.**, Cyclic ADP-ribose activation of the ryanodine receptor is mediated by calmodulin, *Nature*, 370, 307, 1994.
58. **Morrissette, J., Heisermann, G., Cleary, J., Ruoho, A., and Coronado, R.,** Cyclic ADP-ribose induced calcium release in rabbit skeletal muscle sarcoplasmic reticulum, *FEBS Lett.*, 330, 270, 1993.
59. **Giannini, G., Clementi, E., Ceci, R., Marziali, G., and Sorrentino, V.,** Expression of a ryanodine receptor-Ca^{2+} channel that is regulated by TGF-β, *Science*, 257, 91, 1992.
60. **Brillantes, A.-M. B.,** Stabilization of calcium release channel (ryanodine receptor) function by FK506-binding protein, *Cell*, 77, 513, 1994.
61. **Berridge, M. J.,** A tale of two messengers, *Nature*, 365, 388, 1993.
62. **Koshiyama, H., Lee, H., and Tashjian, A.,** Novel mechanism of intracellular calcium release in pituitary cells, *J. Biol. Chem.*, 266, 16985, 1991.
63. **Currie, K., Swann, K., Galione, A., and Scott, R.,** Activation of Ca^{2+}dependent currents in cultured rat dorsal root ganglion neurons by a sperm factor and cyclic ADP-ribose, *Mol. Biol. Cell*, 3, 1415, 1992.
64. **Takasawa, S., Nata, K., Yonekura, H., and Okamoto, H.,** Cyclic ADP-ribose in insulin secretion from pancreatic β cells, *Science*, 259, 370, 1993.
65. **White, A. M., Watson, S. P., and Galione, A.,** Cyclic ADP-ribose-induced Ca^{2+} release from rat brain microsomes, *FEBS Lett.*, 318, 259, 1993.

66. **Hua, S.-Y., Tokimasa, T., Takasawa, S., Furuya, Y., Nohmi, M., Okamoto, H., and Kuba, K.,** Cyclic ADP-ribose modulates Ca^{2+} release channels for activation by physiological Ca^{2+} entry in bullfrog sympathetic neurons, *Neuron*, 12, 1073, 1994.
67. **Whalley, T., McDougall, A., Crossley, I., Swann, K., and Whitaker, M.,** Internal calcium release and activation of sea urchin eggs by cGMP are independent of the phosphoinositide signaling pathway, *Mol. Biol. Cell*, 3, 373, 1992.
68. **Islam, M. S., Larsson, O., and Berggren, P.-O.,** Cyclic ADP-ribose in β cells, *Science*, 262, 584, 1993.
69. **Rutter, G. A., Theler, J.-M., Li, G., and Wollheim, C. B.,** Ca^{2+} stores in insulin-secreting cells: lack of effect of cADP ribose, *Cell Calcium*, 16, 71, 1994.
70. **Lee, H. C.,** A signaling pathway involving cyclic ADP-ribose, cGMP and nitric oxide, *News Physiol. Sci.*, 9, 134, 1994.

Chapter 5

RYANODINE RECEPTOR-SPECIFIC SCORPION TOXINS: NEW PROBES FOR THE ANALYSIS OF CHANNEL STRUCTURE AND FUNCTION

Jeffery Morrissette and Roberto Coronado

TABLE OF CONTENTS

1. INTRODUCTION

Ryanodine receptors are intracellular Ca^{2+} channels which in muscle are ultimately responsible for the initiation and propagation of intracellular Ca^{2+} signals.[1] In cardiac muscle, Ca^{2+} entry into cells directly triggers the opening of ryanodine receptors which results in release of Ca^{2+} from the main storage site, the sarcoplasmic reticulum (SR).[2] In skeletal muscle, membrane depolarization opens the ryanodine receptor without the participation of the Ca^{2+} current.[3] Other intracellular channels such as IP_3 receptors have been implicated in stimulus-Ca^{2+} release coupling in many types of muscle and non-muscle cells.[4] The presence of IP_3 receptors and ryanodine receptors in the same tissue has also been reported.[5,6] In myocardium, the concerted action of IP_3 and ryanodine receptors could explain the α_1 sympathomimetic positive inotropic effect.[7]

0-8493-8543-1/95/$0.00+$.50

The availability of pharmacological agents to selectively block or enhance the activity of a specific intracellular Ca^{2+} channel seems critical to evaluate its role in signal transduction. For the case of ryanodine receptors, the plant alkaloid, ryanodine, is the only agent that is sufficiently specific to permit inferences about the channel's contribution to Ca^{2+} signals *in situ*.[8] The usefulness of ryanodine is limited, however, by its slow association and dissociation kinetics, which make the onset of the pharmacological effect slow and essentially irreversible.[9] Furthermore, the effects of ryanodine are complex, in that there exists both high and low affinity binding sites for the alkaloid on the receptor. Ryanodine binding to the high affinity site opens the channel to a low conductance state, while subsequent binding to the low affinity site results in channel closure.[10] By depleting the SR of stored Ca^{2+}, ryanodine may also blunt Ca^{2+} signals initiated by non-ryanodine receptor pathways or channels.

Many studies have indicated that scorpion venoms are a rich source of peptides highly specific for a single type of channel.[11,12] We thus turned to these venoms in search for a ligand specific for ryanodine receptors that could act quickly, reversibly, and in a simple manner. As a screening technique we used rabbit skeletal muscle SR enriched in terminal cisternae and [^{3}H]ryanodine in a radioligand binding assay under nonequilibrium conditions. Ryanodine binds to the receptor irreversibly and preferentially to the open state of the channel.[13] Compounds such as caffeine and adenine nucleotides, which activate the channel, increase the number of tagged receptors per unit time, whereas inhibitors such as Mg^{2+} and ruthenium red decrease the number of tagged receptors per unit time.[14] Thus, by measuring [^{3}H]ryanodine binding activity at a time (90 min at 37°C) before equilibrium is reached, we could correlate changes in binding activity of [^{3}H]ryanodine with changes in the number of open channels produced by a given venom fraction. We found that the venom of the North African scorpion *Buthotus hottentota* stimulated [^{3}H]ryanodine binding to rabbit skeletal muscle SR in a dose-dependent manner. Furthermore, single-channel recordings comfirmed that the venom activated Ca^{2+} release channels incorporated into planar bilayers. In this chapter we describe the purification and characterization of ryanodine receptor-specific peptide components of *Buthotus hottentota* venom and discuss their potential uses to elucidate the structure and function of ryanodine receptors.

2. PURIFICATION OF TOXINS

Approximately 100 mg of lyophilized *Buthotus hottentota* venom was resuspended in 3 ml of deionized water and sonicated. The mucus was removed by aspiration, yielding a water-soluble fraction with a protein concentration of about 4 mg/ml. Active toxins were purified from the water-soluble fraction by a total of three rounds of HPLC using preparative and analytical C_{18} reverse phase columns, as described in Figure 1. Figure 1A corresponds to the purification scheme of two stimulatory toxins and Figure 1B corresponds to the purification scheme of two inhibitory toxins, all from the same venom. Bound peptides were eluted with acetonitrile gradients indicated by the dashed lines and monitored by light absorption at 254 nm. After each round of fractionation (HPLC I, II, IIIA, IIIB in Figure 1A; HPLC I, A, B, C in Figure 1B), the largest and best-resolved A_{254} peaks were collected, dried under vacuum centrifugation, and resuspended in a minimal volume of distilled water. The resuspended fractions were tested for activity using the [^{3}H]ryanodine binding assay described below at approximately equivalent concentrations in A_{254} units/ml. Active fractions, indicated by the arrows, were subjected to subsequent rounds of fractionation. The binding activity of each of the numbered HPLC peaks in Figure 1 is given as a percentage of control binding (control = 100%):

HPLC I: peak 1 = 5%, peak 2 = 76%, peak 3 = 438%, peak 4 = 349%, peak 5 = 627%, peak 6 = <1%,
HPLC II: peak 1 = 106%, peak 2 = 96%, peak 3 = 135%, peak 4 = 300%, peak 5 = 437%, peak 6 = 173%,
HPLC IIIA: peak 1 = 90%, peak 2 = 95%, peak 3 = 109%, peak 4 = 453%, peak 5 = 100%,
HPLC IIIB: peak 1 = 107%, peak 2 = 196%, peak 3 = 417%, peak 4 = 566%, peak 5 = 140%,
HPLC A: peak 1 = 93%, peak 2 = 53%, peak 3 = 112%, peak 4 = 89%, peak 5 = 26%, peak 6 = 49%, peak 7 = 8%, peak 8 = 85%, peak 9 = 270%, peak 10 = 300%, peak 11 = 307%, peak 12 = 303%, peak 13 = 108%, peak 14 = 1%, peak 15 = 140%, peak 16 = 98%,
HPLC B: peak 1 = 143%, peak 2 = 83%, peak 3 = 37%,
HPLC C: peak 1 = 91%, peak 2 = 94%, peak 3 = 75%, peak 4 = 91%, peak 5 = <1%, peak 6 = 71%.

Upon fractionation of the *Buthotus* venom we found peaks that, like the whole venom, stimulated [^{3}H]ryanodine binding. However, we also detected peaks which inhibited [^{3}H]ryanodine binding. For example, after the initial preparative HPLC (HPLC I), peak 5 stimulated specific [^{3}H]ryanodine binding approximately 600% over the control, while peak 6 completely inhibited the control binding activity. In two subsequent rounds, we fractionated stimulatory peak 5 from HPLC I into two single peaks labeled $Buth_A$-1 and $Buth_A$-2 (Figure 1A). Similarily, we fractionated the inhibitory peak 6 from HLPC I into two single peaks labeled $Buth_I$-1 and $Buth_I$-2 (Figure 1B). The peptide composition of the four toxin fractions was determined by SDS-PAGE (sodium dodecyl sulfate-polyacrylamide gel electrophoresis) on 6 to 20% polyacrylamide gradient gels. The active toxin fractions consisted of an apparently single polypeptide with an estimated molecular weight of approximately 7, 12, 13, and 13 KDa for $Buth_A$-1, $Buth_A$-2,

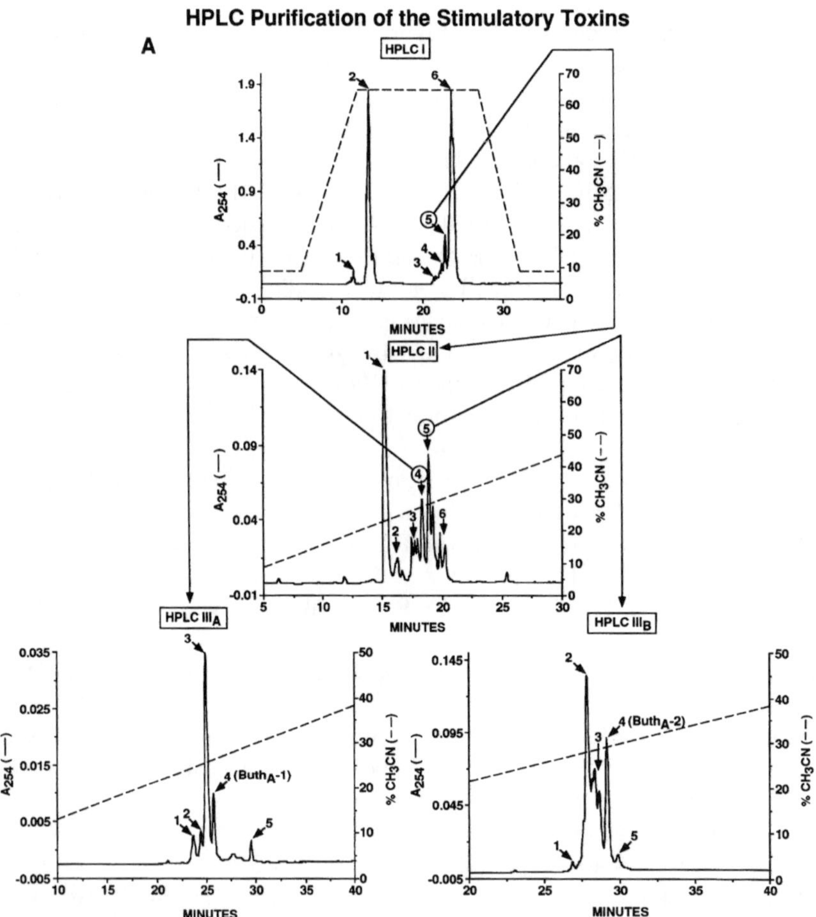

FIGURE 1. Purification scheme for the stimulatory and inhibitory toxins from *Buthotus hottentota* venom. **(A) Top panel:** chromatographic profile of HPLC I resulting from the injection of 500 μg of water soluble venom into a C_{18} reverse phase column (μBondapac-Waters, 0.78 × 30 cm). Fractions were eluted with a linear 8.75 to 65.0% acetonitrile gradient containing 0.1% trifluoroacetic acid run at a flow rate of 1.0 ml/min for 60 min. **Middle panel:** chromatographic profile of HPLC II resulting from the injection of approximately 100 μg of the stimulatory fraction 5 from HPLC I into a C_{18} reverse phase column (Vydac, 0.46 × 25 cm). Fractions were eluted with a linear 8.75 to 65.0% acetonitrile gradient containing 0.1% trifluoroacetic acid run at a flow rate of 1.0 ml/min for 60 min. **Bottom panel:** chromatographic profiles resulting from the injection of approximately 5 μg of the stimulatory fraction 4 from HPLC II (left) or 5 μg of the stimulatory fraction 5 from HPLC II (right) into the Vyadac column. Fractions were eluted with a 8.75 to 42.5% acetonitrile gradient containing 0.1% trifluoroacetic acid run at a flow rate of 1.0 ml/min.

$Buth_I$-1, and $Buth_I$-2, respectively. Furthermore, all four toxin fractions eluted as single symmetrical peaks when further HPLC fractionation was attempted. The percentage of toxin in the whole venom, estimated by the absorbance at 254 nm,

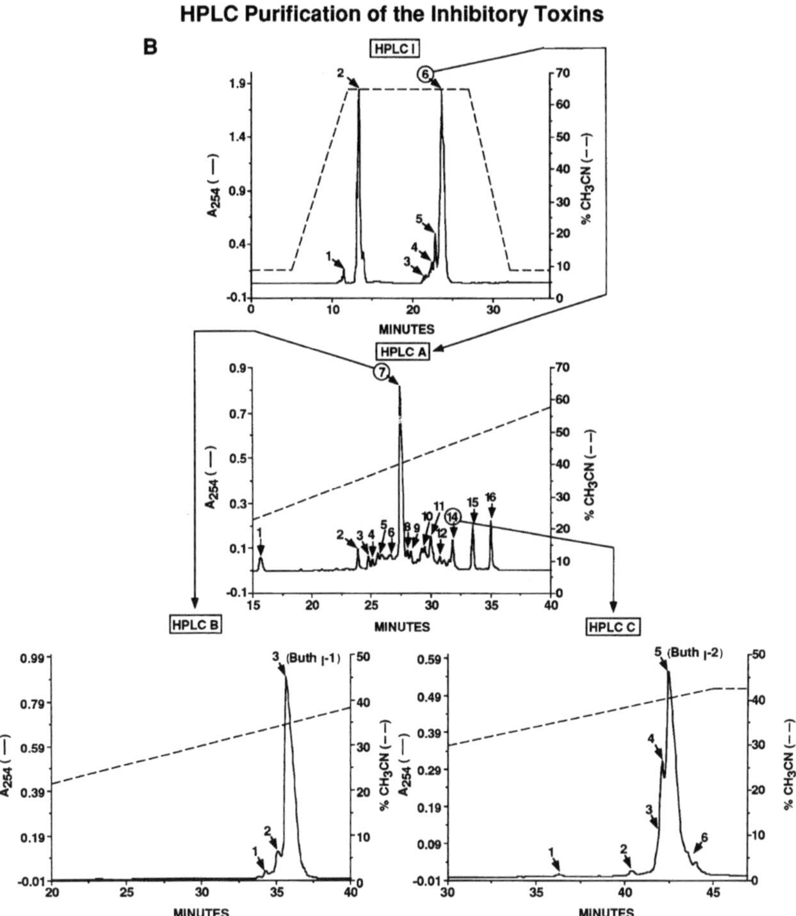

FIGURE 1. (continued) (B) Top panel: chromatographic profile of HPLC I resulting from the injection of approximately 500 μg of the water-soluble venom into a C_{18} reverse phase column (μBondapac-Waters, 0.78 × 30 cm). Fractions were eluted with a linear 8.75 to 65.0% acetonitrile gradient containing 0.1% trifluoroacetic acid run at a flow rate of 1.0 ml/min for 60 min. **Middle panel:** chromatographic profile of HPLC A resulting from the injection of approximately 500 μg of the inhibitory fraction 6 from HPLC I into a C_{18} reverse phase column (Vydac, 0.46 × 25 cm). Fractions were eluted with a linear 8.75 to 65.0% acetonitrile gradient containing 0.1% trifluoroacetic acid run at a flow rate of 1.0 ml/min for 60 min. **Bottom panel:** chromatographic profiles resulting from the injection of approximately 200 μg of the inhibitory fraction 7 from HPLC A (left) or 200 μg of the inhibitory fraction 14 from HPLC A (right) into the Vyadac column. Fractions were eluted with a 8.75 to 42.5% acetonitrile gradient containing 0.1% trifluoroacetic acid run at a flow rate of 1.0 ml/min.

was 0.04% for $Buth_A$-1, 0.35% for $Buth_A$-2, 17.7% for $Buth_I$-1, and 0.63% for $Buth_I$-2. The higher venom content of inhibitory toxins was surprising given that unfractionated venom stimulated [^{3}H]ryanodine binding.

3. CHARACTERIZATION OF TOXINS

3.1. EFFECTS ON [^{3}H]RYANODINE BINDING

[^{3}H]ryanodine binding to rabbit skeletal muscle heavy SR has been described in detail elsewhere.[15] Briefly, HPLC fractions (0.05 to 0.1 A_{254} units/ml), rabbit skeletal muscle heavy SR (40 μg protein), and 7 n*M* [^{3}H]ryanodine were incubated for 90 min at 36°C in 0.1 ml of 0.2 *M* KCl, 1 m*M* Na_2EGTA [ethylene glycol-bis(aminoethyl ether) tetraacetic acid], 0.995 m*M* $CaCl_2$, 10 m*M* Na-PIPES (1,4-piperazine ethanesulfonic acid), at pH 7.2. The calculated free Ca^{2+} concentration, as determined with a computer program using the affinity constants of Fabiato,[16] was 10 μ*M*. Samples were filtered on Whatman GF/B glass fiber filters, washed twice with 5 ml of distilled water, and counted by scintillation. The nonspecific binding was measured in the presence of 10 μ*M* unlabeled ryanodine and was subtracted from each sample. Whole *Buthotus* venom, as well as the purified stimulatory toxins, increased [^{3}H]ryanodine binding in a dose dependent manner. On the other hand, the purified inhibitory toxins decreased [^{3}H]ryanodine binding, also in a dose dependent manner. The concentrations which produced 50% of the maximal stimulation or 50% of the inhibition (ED_{50}) were 70 n*M* and 0.9 μ*M* for the stimulatory toxins $Buth_A$-1 and $Buth_A$-2, respectively, and were 30 μ*M* and 6.9 μ*M* for the inhibitory toxins $Buth_I$-1 and $Buth_I$-2, respectively. The higher affinity of the stimulatory toxins could explain the fact that the whole venom was stimulatory even when inhibitory toxins were present in higher amounts. It is also possible that the inhibitory toxins may be inactivated in the whole venom by components later removed during purification.

In order to determine if the toxins interacted with ryanodine receptors in a specific manner, we performed receptor-radioligand binding assays in muscle and brain microsomes in the presence and absence of *Buthotus* toxins. We investigated the possible interactions of toxins with the dihydropyridine receptor and the voltage-dependent Na^+ channel in skeletal muscle transverse tubule membranes, and with the IP_3 receptor in brain microsomes. Neither the unfractionated *Buthotus* venom nor the purified toxins interfered with the binding of [^{3}H]PN200-110, [^{3}H]saxitoxin, or [^{3}H]IP_3 to their respective receptors. This was significant, because at the same concentration the toxins produced a significant stimulation or inhibition of [^{3}H]ryanodine binding to the skeletal ryanodine receptor. By this criterion, that is, the lack of interference of toxins with other ligand-receptor reactions, the *Buthotus* toxins appeared to be specifically targeted for the ryanodine receptor.

3.2. EFFECTS ON SR Ca^{2+} RELEASE

The physiological consequences of the interaction between the toxins and the ryanodine receptor were investigated by determining the changes produced by the toxins on the Ca^{2+} permeability of SR vesicles. Changes in extravesicular free Ca^{2+} of a microsomal SR suspension were monitored with the Ca^{2+}

indicator dye Fura-2 on a dual excitation wavelength Hitatchi F-2000 fluorescence spectrophotometer.[17] Approximately 150 µg of heavy SR vesicles were suspended in 400 µl of a medium containing 0.1 *M* KCl, 5.0 m*M* phosphocreatine, 5.0 µg/ml creatine phosphokinase, 0.5 µ*M* Fura-2 (free acid), and 20 m*M* HEPES-TRIS pH 7.2. This solution was placed in a quartz cuvette in the spectrophotometer under constant stirring at 30°C. The free Ca^{2+} was in the range of 1 to 3 µ*M* and was present as a contaminant since no Ca^{2+} salt was added to this solution. As shown in Figure 2, SR vesicles were actively loaded with Ca^{2+} by stimulating the Ca^{2+} pump after addition of 1 to 2 m*M* MgATP to the cuvette. Ca^{2+} loading of SR vesicles was inferred from the decrease in extravesicular Ca^{2+} following the addition of ATP. Ca^{2+} release from the SR was inferred from the increase in extravesicular free Ca^{2+} stimulated by the addition of the ryanodine receptor activator, caffeine, or by the addition of stimulatory *Buthotus* toxins. In the control run shown in Figure 2A, approximately 500 n*M* Ca^{2+} was released upon the addition of 10 m*M* caffeine to Ca^{2+}-loaded SR. The caffeine-induced release could be blocked by the previous addition to the SR suspension of the ryanodine receptor blocker, ruthenium red (Figure 2B). In a similar manner, Figures 2C and 2D showed that addition of either of the purified inhibitory toxins to the Ca^{2+}-loaded SR suspension prevented the release induced by caffeine. These results indicated that $Buth_I$-1 and $Buth_I$-2, like ruthenium red, were blocking the channel, thus preventing the activation of the channel by caffeine. This interpretation was consistent with that obtained from [^{3}H]ryanodine binding, in that the toxins seemed to stabilize the channels in a closed conformation. On the other hand, Figures 2E and 2F showed that the stimulatory toxins, like caffeine, caused the release of 300 to 600 n*M* Ca^{2+} following ATP-dependent uptake. The release was mediated by ryanodine receptors since the previous addition of ruthenium red blocked the toxin-induced release (Figures 2G and 2H). These results indicated that $Buth_A$-1 and $Buth_A$-2 stabilized the channel in an open conformation, thus stimulating the release of stored Ca^{2+}. A toxin-induced opening of the channel was also consistent with the stimulation of [^{3}H]ryanodine binding described above.

3.3. EFFECTS OF INHIBITORY TOXINS ON SINGLE CHANNEL ACTIVITY

The interactions of the most abundant toxins in *Buthotus* venom ($Buth_I$-1 and $Buth_I$-2) with the ryanodine receptor were confirmed by planar bilayer recordings. Bilayer formation and recording were performed as previously described.[18] Briefly, lipid bilayers were cast using a 1:1 ratio of brain phosphatidylethanolamine and brain phosphatidylserine dissolved in decane at a concentration of 20 mg/ml. The lipid solution was spread over a 300 µ*M* diameter aperature separating two aqueous chambers designated cis and trans. SR vesicles were added to the cis chamber containing 250 m*M* CsCl, 10 m*M* HEPES-TRIS pH 7.2. The trans chamber contained 50 m*M* CsCl, 10 m*M* HEPES-TRIS pH 7.2. The cis chamber was connected via an Ag/AgCl electrode to the headstage input of a List L/M EPC

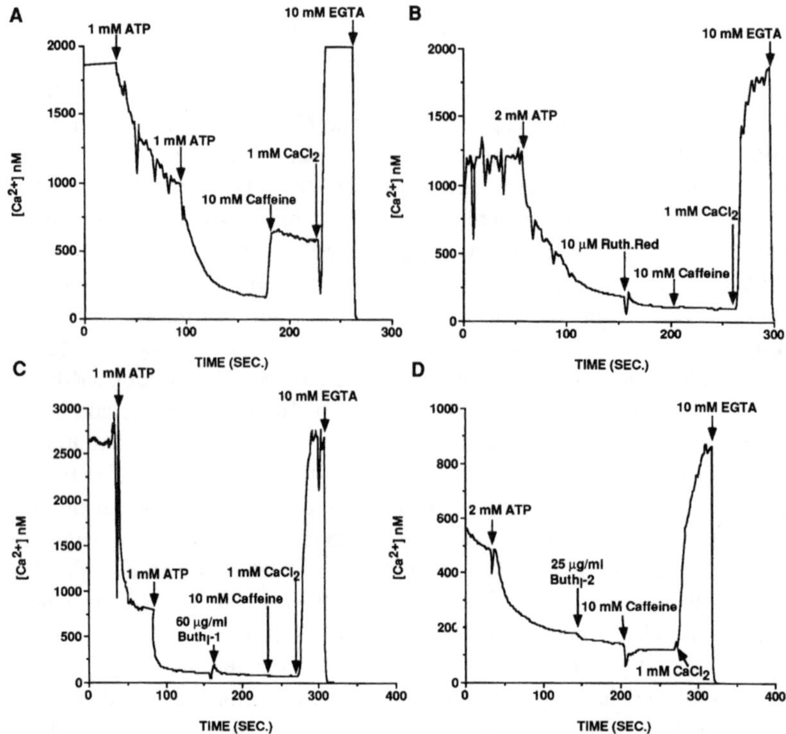

FIGURE 2. Effect of the toxins on Ca^{2+} release from SR vesicles. (**A**) Ca^{2+} release induced by 10 m*M* caffeine, time of the addition is indicated by the arrow. (**B**) The ryanodine receptor antagonist ruthenium red blocks the caffeine-induced Ca^{2+} release. (**C**) and (**D**) The inhibitory toxins, $Buth_I$-1 and $Buth_I$-2, block the caffeine-induced release. (**E**) and (**F**) The stimulatory toxins, $Buth_A$-1 and $Buth_A$-2, induced a release of stored Ca^{2+}. (**G**) and (**H**) Ruthenium red blocked the toxin-induced release. $CaCl^2$ (1 m*M*) and Na_2EGTA (10 m*M*) were added at the end of each experiment in order to determine F_{max} and F_{min}, which are necessary to calculate free Ca^{2+} according to the ratiometric technique.[17]

7 amplifier while the trans chamber was held at ground. All recordings were taken at 10 mV, filtered at 1 kHz, and digitalized at 3 kHz for analysis. The major effect of $Buth_I$-1 was to decrease the number of open events, thus resulting in a decrease in the channel P_o (open probability) from 0.42 to 0.04. It was apparent that $Buth_I$-1 produced a constant inhibition of the channel that was not relieved with time. Similar results were obtained for a single channel recorded in the presence of $Buth_I$-2. The P_o decreased from 0.1 to 0.026 following the addition of $Buth_I$-2. As with $Buth_I$-1, the inhibition produced by $Buth_I$-2 was associated with a decrease in the number of open events per time and was persistent throughout the recording. These results clearly showed that the inhibitory toxins produced a closure of the channel which, in turn, explained the inhibition of caffeine-induced release of SR-stored Ca^{2+} and the inhibition of the binding of $[^3H]$ryanodine to the receptor.

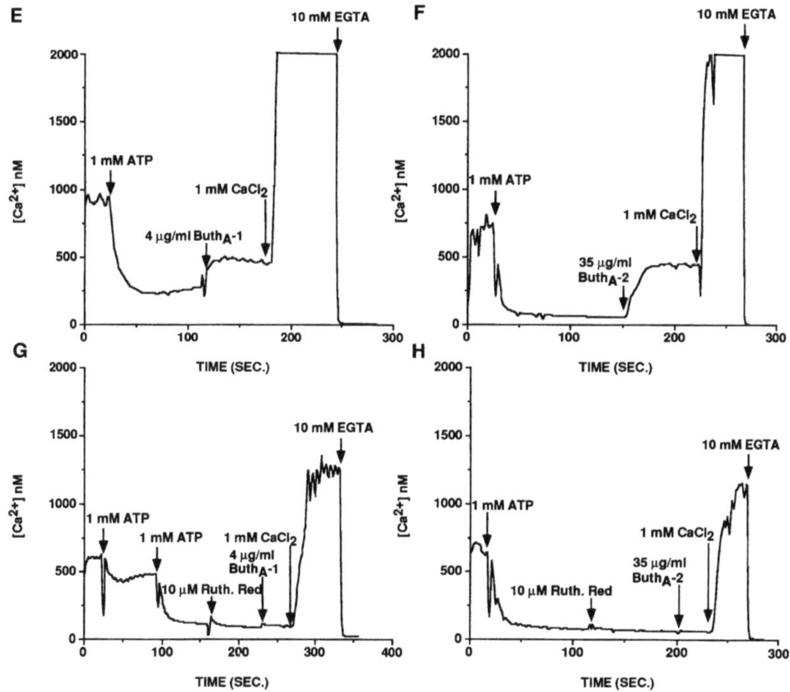

FIGURE 2. (continued)

4. POTENTIAL USES OF TOXINS

Given the intracellular location of ryanodine receptors, the elucidation of the mechanism of Ca^{2+} release from intracellular stores depends critically on the specificity of pharmacological agents to selectively alter a single intracellular Ca^{2+} channel type. We now know that a small fraction of scorpion venoms have peptide toxins that appear to be targeted specifically against intracellular Ca^{2+} channels and are present in these venoms in biochemically relevant quantities.[15,19] These toxins should be invaluable in relating structure to function of this channel protein and should be useful in understanding how intracellular Ca^{2+} signals are generated.

4.1. PHARMACOLOGICAL DISSECTION OF INTRACELLULAR Ca^{2+} RELEASE

Two Ca^{2+} pools, controlled by an IP_3-sensitive and a Ca^{2+}-sensitive channel, have been shown to operate during Ca^{2+} transients in secretory and nonexcitable cells.[20] Whether two Ca^{2+} pools are also activated in striated muscle during stimulus-contraction coupling is not known. However, this is an attractive model since the rate of Ca^{2+}-induced Ca^{2+} release and the rate of IP_3-induced Ca^{2+} release in skeletal muscle membranes turned out to be equally fast.[21]

Interactions between Ca^{2+} pools controlled by IP_3 and ryanodine receptors in the same cell may serve several purposes. For example, it could serve to rapidly amplify a small release of Ca^{2+} produced by IP_3 into a much larger elevation of myoplasmic Ca^{2+} produced by Ca^{2+}-dependent activation of ryanodine receptors. The activation of phospholipase C (PLC) by Ca^{2+} [22] also makes possible an alternative cascade, in which the transient increase in Ca^{2+} following voltage-activated Ca^{2+} release would activate PLC and production of IP_3. There are numerous questions that may be addressed using ryanodine-receptor specific toxins. The most pressing ones seem to be (1) which release pathway, IP_3 receptor, ryanodine receptor, or both, is involved in Ca^{2+} waveforms in cells that have both types of channels, and (2) since both receptors are known to be Ca^{2+}-sensitive,[23,24] is the activation of one Ca^{2+} pool subsequently involved in the activation of the other, thus resulting in synergistic Ca^{2+} release? With the availability of ryanodine receptor-specific toxins, the ryanodine-sensitive Ca^{2+} pool can be selectively depleted with stimulatory toxins or blocked with inhibitory toxins such that it is should be possible to determine its relationship to the IP_3 sensitive Ca^{2+} pool.

4.2. MAPPING OF REGULATORY DOMAINS OF RYANODINE RECEPTORS

The large size of the ryanodine receptor (≈565 kDa) makes the task of identifying domains of the receptor that are important for channel function quite difficult. Scorpion toxins can be used to identify the areas of the receptor critical to channel function. Since the *Buthotus* toxins affect channel gating, it seems likely that the binding sites for the toxins could be physically close to domains critically involved in the opening and closing of the channel. One possible approach here is the use of the toxins as potential substrates for the synthesis of photoaffinity labels. Once binding sites are identified, subsequent site-directed mutagenesis of these domains should provide valuable information of the specific amino acids involved in channel gating.

4.3. IDENTIFICATION OF RYANODINE RECEPTOR ISOFORMS

Three different genes (RYR-1, RYR-2, and RYR-3) encoding for a skeletal, cardiac, and brain isoform of the ryanodine receptor, respectively, have been identified.[1,3] However the expression of the three isoforms is not completely tissue specific. The receptor encoded by RYR-1 was found to be the predominant form expressed in skeletal muscle, but was also present in mouse cerebellum.[25] Likewise, the receptor encoded by RYR-2 was the predominant form expressed in the heart, but was also the predominant form expressed in the brain.[1,26] Also found in the brain, as well as in smooth muscle, was the receptor encoded by RYR-3.[27] The cardiac receptor (RYR-2) and the brain receptor (RYR-3) were found to be 66% and 70% homologous to the skeletal receptor (RYR-1).[26,27] Although there is a great deal of homology between the receptor isoforms, the skeletal and cardiac receptors are functionally different in terms of channel conductance, sensitivity to activation by Ca^{2+} and caffeine, and inhibition by Mg^{2+}.[28] Furthermore, the cardiac and skeletal isoforms seem to be activated by different mechanisms during excitation-contraction coupling. In cardiac muscle, a small influx of Ca^{2+} across the sarcolemal membrane triggers a much more massive Ca^{2+} release via the ryanodine receptor.[2] In skeletal muscle, activation of the sarcolemmal dihydropyridine receptor by membrane depolarization leads to Ca^{2+} release via the ryanodine receptor, possibly through a mechanical coupling between the dihydropyridine and ryanodine receptors.[3]

Scorpion toxins that are isoform-specific should be quite useful for identification and characterization of ryanodine receptor isoforms. Valdivia et al. have found a peptide toxin from the scorpion *Pandinus imperator*, named imperatoxin activator (IpTxa), that stimulates [^{3}H]ryanodine binding and single-channel activity of skeletal but not cardiac receptors.[19] The stimulatory toxins from *Buthotus,* described here, did not differentiate between receptor isoforms. Thus, a comparison between the binding site for IpTxa and those of $Buth_A$-1 and $Buth_A$-2 may provide insight into the regions of the ryanodine receptor which confer the functional differences between receptor isoforms. Furthermore, isoform-specific toxins may prove useful in the selective purification of ryanodine receptor isoforms from tissues such as the brain, that contain more than a single ryanodine receptor isoform.

ACKNOWLEDGMENTS

Supported by National Institutes of Health grant GM 36852, and by grants from the American Heart Association and Muscular Dystrophy Association of America.

REFERENCES

1. **Coronado, R., Morrissette, J., Sukhareva, M., and Vaughan, D. M. V.,** Invited review: Structure and function of ryanodine receptors, *Am. J. Physiol.*, 266, C1485, 1994.
2. **Stern, M. D. and Lakatta, E. G.,** Excitation-contraction coupling in the heart: the state of the question, *FASEB J.,* 6, 3029, 1992.
3. **McPherson, P. S. and Campbell, K. P.,** The ryanodine receptor/Ca^{2+} release channel, *J. Biol. Chem.*, 268, 13765, 1993.
4. **Mikoshiba, K.,** Inositol 1,4,5-trisphosphate receptor, *Trends Pharmacol. Sci.*, 14, 86, 1993.
5. **Jorgensen, A. O., Shen, A. C.-Y., Arnold, W., McPherson, P. S., and Campbell, K. P.,** The Ca^{2+}-release channel/ryanodine receptor is localized in junctional and corbular sarcoplasmic reticulum in cardiac muscle, *J. Cell Biol.,* 120, 969, 1993.
6. **Moschella, M. C. and Marks, A. R.,** Inositol 1,4,5-trisphosphate receptor expression in cardiac myocytes, *J. Cell Biol.*, 120, 1137, 1993.

7. **Otani, H., Otani, H., and Das, D.,** a_1-Adrenoceptor-mediated phosphoinositide breakdown and inotropic response in rat left ventricular papillary muscles, *Circ. Res.,* 62, 8, 1988.
8. **Beuckelmann, D. J. and Wier, W. G.,** Mechanism of release of calcium from sarcoplasmic reticulum of guinea-pig cardiac cells, *J. Physiol. (Lond.),* 405, 233, 1988.
9. **Pessah, I., Francini, A., Scales, D., Waterhouse, A., and Casida, J.,** Calcium-ryanodine receptor complex. Solubilization and partial characterization from skeletal muscle junctional sarcoplasmic reticulum vesicles, *J. Biol. Chem.*, 261, 8643, 1986.
10. **Wang, J. P., Needleman, D. H., and Hamilton, S. L.,** Relationship of low affinity [^{3}H]ryanodine binding sites to high affinity sites on the skeletal muscle Ca^{2+} release channel, *J. Biol. Chem.*, 268, 20974, 1993.
11. **Strichartz, G., Rando, T., and Wang, G. K.,** An integrated view of the molecular toxicology of sodium channel gating in excitable cells, *Annu. Rev. Neurosci.*, 10, 237, 1987.
12. **Miller, C., Moczydlowski, E., Latorre, R., and Phillips, M.,** Charybdotoxin, a protein inhibitor of single Ca^{2+}-activated K+ channels from mammalian skeletal muscle, *Nature*, 313, 316, 1985.
13. **Imagawa, T., Smith, J. S., Coronado, R., and Campbell, K. P.,** Purified ryanodine receptor from muscle sarcoplasmic reticulum is the Ca^{2+}-permeable pore of the calcium release channel, *J. Biol. Chem.*, 262, 16636, 1987.
14. **Pessah, I. N., Stambuk, R. A., and Casida, J. E.,** Ca^{2+}-activated ryanodine binding: mechanism of sensitivity and intensity modulation by Mg^{2+}, caffeine, and adenine nucleotides, *Mol. Pharmacol.*, 31, 232, 1987.
15. **Valdivia, H. H., Fuentes, O., El-Hayek, R., Morrissette, J., and Coronado, R.,** Activation of the ryanodine receptor Ca^{2+} release channel of sarcoplasmic reticulum by a novel scorpion venom, *J. Biol. Chem.*, 266, 19135, 1991.
16. **Fabiato, A.,** Computer programs for calculating total from specified free or free from specified total ionic concentrations in aqueous solutions containing multiple metals and ligands, *Meth. Enzymol.*, 157, 387, 1988.
17. **Grynkiewicz, G., Poenie, M., and Tsien, R. Y.,** A new generation of Ca^{2+} indicators with greatly improved fluorescence properties, *J. Biol. Chem.*, 260, 3440, 1985.
18. **Coronado, R., Kawano, S., Lee, C. J., Valdivia, C., and Valdivia, H. H.,** Planar bilayer recording of ryanodine receptors of sarcoplasmic reticulum, *Meth. Enzymol.*, 207, 699, 1992.
19. **Valdivia, H. H., Kirby, M. S., Lederer, W. J., and Coronado, R.,** Scorpion toxins targeted against the sarcoplasmic reticulum Ca^{2+}-release channel of skeletal and cardiac muscle, *Proc. Natl. Acad. Sci. U.S.A.*, 89, 12185, 1992.
20. **Connor, J. A.,** Intracellular calcium mobilization by inositol 1,4,5-trisphosphate: Intracellular movements and compartmentalization, *Cell Calcium*, 14, 185, 1993.
21. **Valdivia, C., Vaughan, D., Potter, B. V. L., and Coronado, R.,** Fast release of $^{45}Ca^{2+}$ induced by inositol 1,4,5-trisphosphate and Ca^{2+} in the sarcoplasmic reticulum of rabbit skeletal muscle. Evidence for two types of Ca^{2+} release channels, *Biophys. J.*, 61, 1184, 1992.
22. **Baird, J. G. and Nahorski S. R.,** Increased intracellular calcium stimulates ^{3}H-inositol polyphosphate accumulation in rat cerebral cortical slices, *J. Neurochem.*, 54, 555, 1990.
23. **Finch, E. A., Turner, T. J., and Goldin, S. M.,** Calcium as a coagonist of inositol 1,4,5-triphosphate-induced calcium release, *Science*, 252, 443, 1991.
24. **Smith, J. S., Coronado, R., and Meissner, G.,** Single channel measurements of the calcium release channel from skeletal muscle sarcoplasmic reticulum, *J. Gen. Physiol.*, 88, 573, 1986.
25. **Kuwajima, G. A., Futatsugi, A., Niinobe, M., Nakanishi, S., and Mikoshiba, K.,** Two types of ryanodine receptors in mouse brain: skeletal muscle type exclusively in purkinje cells and cardiac muscle type in various neurons, *Neuron*, 9, 1133, 1992.

26. **Otsu, K., Willard, H. F., Khanna, V. K., Zorzato, F., Green, N. M., and Maclennan, D. H.,** Molecular cloning of cDNA encoding the Ca^{2+} release channel (ryanodine receptor) of rabbit cardiac muscle sarcoplasmic reticulum, *J. Biol. Chem.*, 265, 13472, 1990.
27. **Hakamata, Y., Nakai, J., Takeshima, H., and Iomoto, K.,** Primary structure and distribution of a novel ryanodine receptor/calcium release channel from rabbit brain, *FEBS Lett.*, 312, 229, 1992.
28. **Zimanyi, I. and Pessah, I. N.,** Comparison of [^{3}H]ryanodine receptors and Ca++ release from rat cardiac and rabbit skeletal muscle sarcoplasmic reticulum, *J. Pharmacol. Exp. Ther.*, 256, 938, 1991.

Chapter 6

MOLECULAR BIOLOGY OF RYANODINE RECEPTORS

Vincenzo Sorrentino

TABLE OF CONTENTS

1. INTRODUCTION

The application of molecular biological methods has greatly improved our knowledge of ryanodine receptors, allowing the cloning of the cDNA corresponding to the mRNAs that code for the known skeletal (RyR1) and cardiac (RyR2) ryanodine receptors, and the unexpected identification of a third mRNA corresponding to a novel isoform (RyR3). Thus, in vertebrates, the family of calcium release channels sensitive to ryanodine comprises three different isoforms.[1,2]

The RyR family, together with the other known family of calcium release channels operated by inositol 1,4,5-trisphosphate ($InsP_3$),[3] and therefore known as the $InsP_3$ receptors ($InsP_3$-Rs), can be grouped to form a larger superfamily of intracellular calcium release channels. In fact, RyRs and $InsP_3$-Rs share a

0-8493-8543-1/95/$0.00+$.50

low, but evident homology in the amino acid sequence that can be observed over the entire length of these proteins. Even more striking than the amino acid sequence homology is the similarity that the two channels show in their three-dimensional architecture.

Both $InsP_3$-R and RyR calcium channels have been proved to be built as tetramers, made of monomers of about 5000 amino acids for the RyRs and 2700 amino acids for the $InsP_3$-Rs. The actual calcium gating channels are formed by the transmembrane domains located in the COOH terminus of each monomer. Coexpression of different isoforms of RyRs in cells such as neurons, skeletal muscle fibers, germ cells, etc., suggests the possibility that some channels may be heterodimeric in their composition.

2. cDNA CLONING AND CHROMOSOMAL LOCALIZATION

cDNAs for mammalian RyR1, i.e., the skeletal muscle isoform, have been cloned from rabbit, human, and pig.[4-7] The rabbit RyR1 mRNA is 15,368 nucleotides long,[5] with an open reading frame (ORF) which encodes a protein of 5037 amino acids with a predicted molecular weight of 565,223. The termination codon is followed by a short 3′ untranslated region (UTR) of 119 nucleotides, while the initiator methionine is preceded by a 138 nucleotides long 5′ untranslated region. A shorter RyR1 transcript of 2400 bp that hybridizes with the 3′ region of the RyR1 has been observed in rabbit brain.[8] This transcript could encode a protein starting at the ATG encoding met^{4382}, yielding a protein of 656 amino acid residues corresponding to the carboxy terminal part of the RyR1 protein.

The full-length RyR1 cDNA has been expressed in CHO and in COS cells, and found to form a channel with calcium responses to caffeine and ryanodine similar to those observed in skeletal muscle.[5,9,10] The RyR1 protein has also been produced in *Spodoptera frugiperda* insect cells by expressing the RyR1 cDNA with a baculivirus-based system.[11] Importantly, expression of RyR1 from baculivirus vectors has been instrumental in proving the significance for the channel properties of RyR1 of the FKBP12 protein, known to associate with RyR in mammalian cells (see Chapter 2).

The RyR2 cDNA, encoding the cardiac isoform, has been cloned from rabbit.[12,13] The rabbit RyR2 mRNA is 16,532 nucleotides long;[12] the initiator methionine is preceded by a 5′ UTR of 307 nucleotides, the ORF codes for a protein of 4969 amino acids with a molecular weight of 564,711. The RyR2 mRNA contains a 1318 nucleotide long 3′ UTR. The RyR2 cDNA, once expressed in *Xenopus* oocytes, was able to induce functional calcium release channels sensitive to caffeine.

cDNAs corresponding to a third isoform, RyR3, have been isolated from mink epithelial cells and from rabbit brain.[14-16] The rabbit RyR3 mRNA is 15,541 nucleotides long.[15] It encodes a protein of 4872 amino acids with a molecular weight of 551,901. The ORF is preceded by a short 5′ UTR and is followed by a 856 bp 3′ UTR. Comparison of mink and rabbit RyR3 mRNA indicates that in mink Mv1Lu cells a cryptic termination site is utilized, resulting in a shorter 3′ UTR. RyR3 is expressed in several tissues, and until now it has not been found to be predominantly expressed in a specific organ or cell type.[14,17] Therefore, we refer to this isoform as RyR3, and not as the brain ryanodine receptor, to avoid confusion with RyR2, that is expressed in brain at higher levels than RyR3.[17,18]

The three RyRs genes are located on different chromosomes. The RYR1 gene has been mapped to chromosome 19q13.1,[19,20] the RYR2 to chromosome 1,[13,21] and RYR3 to the 15q14-q15 region of the human genome.[22] The murine homologues of RYR1, RYR2, and RYR3 have been mapped to the A2-A3 region of chromosome 7, the A1-A2 region of chromosome 13, and the E5-F3 of chromosome 2, respectively.[23,24]

No direct experimental data have been reported yet on the regulatory regions of RyRs. The genomic sequence upstream of the start site of both RyR1 and RyR2 contains two copies of the GC box sequence in both genes.[5,12] A CAAT box is found between the GC boxes in RyR1, but not in RyR2. Neither promoter regions contains a TATA box. Induction of RyR3 mRNA in the Mv1Lu cells was found to be transcriptionally regulated by TGFß,[14] but no details of the 5′ regulatory region of this gene have yet been reported.

Mice with a targeted mutation in the skeletal ryanodine (RYR1) gene have been produced using a replacement vector, where a gene conferring resistance to neomycin was inserted in exon 2 of RYR1 so that, following transfection of this plasmid in ES cells, homologous recombination would allow the interruption of the RyR1 coding frame.[25] Mice carrying one copy of the altered RYR1 allele ($+/skrr^{m1}$) were generated from blastocysts injected with recombinant ES cells. Crosses between heterozygous $+/skrr^{m1}$ mice allow the generation of mice with a nonfunctional RYR1 locus. These mice were able to survive fetal development but died at birth, probably because of respiratory failure due to a functional deficiency of skeletal muscle. Muscle fibers from mutant mice failed to show a contractile response following electrical stimulation, indicating a basic role of RyR1 in excitation-contraction coupling in skeletal muscle.[26] Interestingly, caffeine was still able to induce release from sarcoplasmic reticulum. This could be due to the presence in skeletal muscle of a low amount of RyR3.[14,17]

RyRs have also been documented in birds, fish, and frogs.[27] In skeletal muscle preparations from these species, two Ry-binding proteins, referred to as α and β, have been described.[28] Expression of a cardiac-specific subtype, similar to the mammalian RyR2 protein, has also been observed in chicken heart.[29] On the basis of their biochemical and immunological properties the two proteins expressed in chicken skeletal muscle have been shown to differ from the RyR2 expressed in heart.[30,31] RyR isoforms have also been observed in the avian central nervous system.[32,33]

In chicken, α and β subtypes are differently expressed during muscle development, with the α subtype being expressed at day 10 and the β subtype first appearing at day 15 of embryonic development.[34] Failure to express the α subtype appears to be tightly associated with the crooked neck Dwarf (cn) mutation in chicken.[35,36] No substantial variation in the expression of the β subtype was found between control and cn mutant chicken.

Recently, two cDNAs isolated from frog skeletal muscle have been found to correspond to the mammalian RyR1 and RyR3. Peptide analysis of frog α and β RyRs has shown correspondence of the former with RyR1 and the latter with RyR3 cDNA sequences, respectively.[37] Similar results have been obtained for chicken skeletal muscle. Analysis of the nucleotide sequences of two clones isolated from the chicken skeletal muscle cDNA library indicates that they correspond to mammalian RYR1 and RYR3. Antibodies developed against the chicken homologues of RyR1 recognize the α isoforms, while antibodies against the RyR3 chicken homologues recognize the β subtype.[38]

A gene encoding a *Drosophila* homologue of vertebrate RyRs has been cloned.[39] This gene encodes a protein with about 46% identity with mammalian RyRs. In adult flies, the *Drosophila* RyR gene is expressed in tubular muscles and neuronal tissues. Transcripts of a *Drosophila* RyR, probably similar to the former one, have been detected in the mesoderm of early stage-9 *Drosophila* embryos and later on in somatic muscles.[40] It is interesting to note that the exon organization of the *Drosophila* gene is much simpler than that of the mammalian one, since there are 27 exons in *Drosophila* while there are approximately 100 exons distributed over about 200 kb in mammalian RyRs (see Chapter 9).

RyRs have also been purified from lobster[41] and in *C. elegans*.[42] Despite some functional properties that are comparable to those of vertebrate RyRs, they show significant differences in the degree of sensitivity to agonists and antagonists of channel activity. With the exception of a very short sequence isolated from a *C. elegans* gene library,[43] no complete sequence is available at the moment from these species.

3. RyR GENE EXPRESSION AND SPLICING

Cells with intracellular calcium-release pools regulated by channels with properties similar to those of RyRs of muscles are present not only in the brain, but also in peripheral tissues.[44,45] RyR1 mRNA was initially described in skeletal muscle[5,6] and RyR2 mRNA was observed in heart, brain, and stomach.[12,13] mRNA for RyR3 was found to be widely expressed in mink and rabbit tissues.[14,15] More recently, all three RyR mRNAs were reported to be expressed in the central nervous system, as well as in almost all tissues analyzed.[17,84]

Expression of RyR isoforms in skeletal muscles of lower vertebrates has shown that, while most muscles express both α and β isoforms (i.e., RyR1 and RyR3), in a few muscles only the α isoform can be detected.[46] Among reptiles, crocodiles and turtles express both α and β isoforms, while lizards and snakes express predominantly the α isoform. In birds, expression of the α isoform alone is mostly confined to the extraocular muscle, one of the fastest contracting muscle in vertebrates, while α and β isoforms are present in the other muscles. The two isoforms expressed in chicken skeletal muscle have been shown to have different regulatory properties.[47]

Extensive characterization of human RyR1 transcripts has indicated that different RyR1 mRNAs exist as the result of a process of splicing that operates at two sites in the RyR1 coding sequence. The first splicing involves a 15 bp sequence encoding 5 amino acids (AGDIQ) after K^{3480}.[5,48] The second splicing involves 18 bp encoding the 6 aa INRQN following V^{3865}.[48] The same sites are also subjected to splicing in the RyR3 transcript, where also a third site undergoing alternative splicing has been observed.[16] Searching for the presence in RyR1 mRNA of the 15 base pair insertion, corresponding to the abovementioned first splicing site in RNA extracted from rabbit slow and fast skeletal muscles, has shown that the two types of transcripts coexist in both types of muscles.[49] In the rabbit RyR2 cDNA, the nucleotide residues 11,146-11,169, coding for the VTGSQRSK amino acids sequence following E^{3715}, were present only in some clones but not in others, a situation likely to result from RNA splicing of the RyR2 transcript.[12]

In the mink Mv1Lu cell line, a splice variant was described that appeared to represent the majority, i.e., 70%, of the RyR3 transcript in this cell line.[14,16] This transcript completely lacks the region coding for the fifth transmembrane domain (see Section 7), a segment that is also strongly conserved in the $InsP_3$-Rs, and is thus expected to be important to the structure/regulation of these channels. Whether the mRNA lacking the TM5 is translated, and the eventual significance of this variant, are not presently known. Two splicing variants of the dopamine receptor D3 have been described that bear some analogy to this splice variant of RyR3 found in Mv1LU cells.[50] One of the alternative splicing variants of D3 lacks exon 2 and induces a stop codon preventing the translation of the first extracellular loop and of the third TM domain. The second D3 splice variant utilizes an alternative internal acceptor site in the fourth exon and as a consequence the resulting protein would lack part of the second extracellular loop and part of the fifth TM domain. Together, these two splice variants account for half of the D3 transcripts present in brain. However, both splice variants of D3, when expressed in CHO cells, do not induce the expression of proteins capable of ligand binding — raising the possibility that neither of the proteins encoded by these splice variants can fold properly and insert into the plasma membrane. Expression and pharmacological characterization of the normal and of the spliced version of RyR3 proteins would be required in order to understand the biological meaning, if any, of this unexpected splicing variant.

A frog RyR1 cDNA clone containing an 18 bp insertion encoding the aa sequence TIINREQGE following G^{3846} has been detected, while other clones missing this sequence contain instead the amino acid sequence TRE.[37] This would correspond to the second splicing site observed in mammalian RyR1.

Sequence analysis of the RyR gene cloned in *Drosophila* has revealed the presence of two alternative sequences at two sites.[39] The first splicing would result from incorporation of either exon 9 or exon 9′, while the second variation in the *Drosophila* RyR sequence results from the utilization of one of the two splice acceptor sites in exon 13. No evidence on the functional relevance of the differentially spliced versions of both mammalian RyR1 and RyR3 or *Drosophila* RyR is available. In $InsP_3$-Rs several splicing sites have been described, and one of them alters a key phosphorylation site in the molecule.[51-53] It is possible that in RyRs this mechanism also may result in structural modifications that modulate the activity of the channel.

4. REGULATORY/DIVERGENCY REGIONS

The RyRs have very conserved amino acid sequences, with a level of identity of 67% between RYR1 and RYR2, 70% between RYR2 and RYR3, and 66% between RYR1 and RYR3; however, there are at least three regions where amino acid sequences diverge significantly.[54] The first divergent region (D1), between residues 4254 and 4631 of RyR1 (and corresponding residues of RyR2 and RyR3), corresponds to the intraluminal loop between TM3 and TM4 and to a larger cytoplasmic loop between TM4 and TM5 (see Figure 1). In this region the three isoforms present the greatest degree of amino acids differences with less than 30% identity between the different isoforms. A diversification at the level of amino acid sequences between the different isoforms is also observed between residues 1302-1404 (D2) and residues 1872-1923 (D3) in the RyR1 protein sequence and the corresponding regions in RyR2 and RyR3. It is possible that these sites contain regulatory domains responsible for some of the physiological and pharmacological differences observed in the different RyR isoforms.

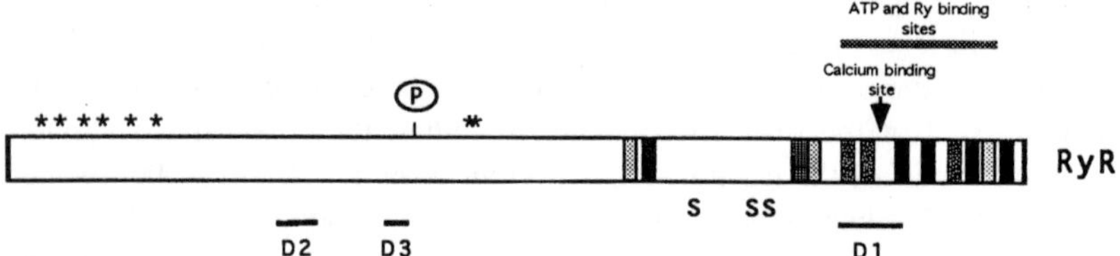

FIGURE 1. Schematic representation of major regulatory sites on the ryanodine receptor. The indicated calcium activation site corresponds to the PEPEPEPE sequence in RyR1. The region likely to correspond to the 76 kDa fragment containing ATP and Ry binding sites is also shown. Asterisks indicate the relative positions of the amino acids found mutated in the RyR1 protein in patients with MH and CCD. Regions containing hydrophobic residues that may potentially traverse the membrane are indicated with bars. They have been filled in black and gray according to the relative hydrophobicity profiles. Divergent regions are labeled D1, D2, and D3. P corresponds to the Ser^{2809} phosphorylated by CaM kinase on RyR2. The three splicing sites identified in mammalian RyRs are each indicated with an S.

It is interesting to note that the D1 amino acid sequences are not strongly conserved across species barriers. Frog and chicken RyR1 and RyR3 share 86% overall homology with mammalian isoforms, and may be as high as 91% in regions that are also conserved among different isoforms. In contrast, homology between avian and mammalian RyR3 in the D1 region is only 57%. Such a relative lack of conservation may apparently contradict the expectation that divergent regions may contain regulatory domains that dictate the functional properties of each isoform. Experimental data, however, support a role of the D1 region, or rather of sequences in this part of the RyR molecule, in modulating calcium release, because antibodies against this part of the RyR1 protein can affect calcium channel function.[55-58]

5. RyR STRUCTURE

Cloning RyR cDNA has provided much information on amino acid composition of these large proteins and has confirmed an evolutionary relationship with the $InsP_3$-Rs. By analogy with the current model for the $InsP_3$-Rs,[3] three regions can be distinguished in the RyR molecule: an amino terminal part including approximately the first 1000 amino acids, which in the $InsP_3$-Rs contains a ligand binding site; a large central part that may form the elaborate scaffold-like structure of the channel as observed by the electron microscope and may act in modulating the calcium-channel activity of the molecule; and thirdly, the last 1000 amino acids that contain the transmembrane domains (TM) involved in forming the calcium pore, and several sites for regulation of channel activity.

No evidence has been obtained yet for a binding region of cADPR on RyRs and at present it is not even clear whether cADPR binds directly to RyRs (see Chapters 3 and 4). Therefore, we have no direct proof that the N-terminal is important for ligand binding or regulation of the channel properties of RyRs. Actually a 76-kDa fragment located in the C-terminal of RyR1 (see Section 6) may contain Ry, ATP, and calcium binding sites. The localization of regulatory sites at the C-terminal of the molecule contrasts with that reported for $InsP_3$-Rs, where it has been directly demonstrated that the 600/800 amino acids at the N-terminal of the $InsP_3$-Rs are important for $InsP_3$ binding.

Evidence in favor of the importance of the N-terminal of RyRs comes from studies in malignant hyperthermia (MH) and in central core disease (CCD), where eight mutations have been identified that cosegregate with the pathological phenotypes in the families where they have been detected, and are absent in the normal population (see Chapter 9). These mutations substitute a Cys for Arg^{163}, Arg for Gly^{248}, Arg for Gly^{341}, Met for Ile^{403}, Ser for Tyr^{522}, Cys for Arg^{614}, Arg for Gly^{2433}, and His for Arg^{2434}. With the exception of His^{2334}Arg and Arg for Gly^{2433}, all the mutations are clustered in a very small region between residues 163 and 615 at the N-terminal of RyR1. All the mutations are localized to areas with considerable sequence conservation between the three isoforms and between different species.

Similar to many other integral membrane proteins, ion channels are difficult to crystallize, thus the structure of RyRs at atomic resolution is not likely to be determined in the near future. However, advances in determining the structure of these molecules have been obtained thanks to electron microscopy.[59] Negative-stain electron microscopy of the purified receptor originally revealed a fourfold symmetric complex (quatrefoils).[60-62] This type of morphology matches the "foot" structure observed, using the electron microscope, in muscle sections (see Chapter 1). The most recent refinement of RyR image reconstruction has provided a few more structural details on these molecules.[63,83] In addition to confirming the existence of a central cavity joined to distinct peripheral cavities by radial channels, these new studies have more clearly demonstrated a structure that is likely to correspond to the transmembrane part of the channel. The estimated dimensions of the molecule indicate that the RyRs have a very large cytoplasmic structure ($27 \times 27 \times 12$ nm) on top of what is likely to be the transmembrane assembly, which extends 7 nm from one face. Electron microscopy has shown that in the cytoplasmic mass of RyR1, up to ten domains can be distinguished in each subunit.[63] These domains are loosely packed together to form a structure that extends out of the organelle membrane like a scaffold. Such a modular organization may allow the RyR channel to undergo conformational changes following interaction with molecules that regulate channel activity. The recent three-dimensional reconstruction of the RyR from skeletal muscle has identified a plug-like mass that lies in the central channel (which potentially corresponds to the transmembrane pore). Radermacher et al.[63] have proposed that this plug-like mass may have a role in mediating channel gating, in response to conformational changes induced by ligand binding to more distant parts of the molecule. Evidence for conformational differences in the RyR1 in its open and closed states has been recently obtained by I. Seysheva, E. Orlova, W. Chiu, M. van Heel, and S. Hamilton (unpublished results). By using EGTA to drive the channel in the closed state, they have been able to obtain a three-dimensional reconstruction of the RyR1 that can be compared with images obtained from similar analysis of specimens of the channel in the open state. As it can be observed in Figure 2, the reconstruction of the channel in its open state reveals a central hole in the transmembrane domain that is apparently absent in the reconstruction of the channel in the closed state. The transition from the open to the closed state also results in a clear structural change in the region located at the corners of the channel.

6. POTENTIAL LIGAND BINDING SITES

Computer analysis of the amino acid sequence of RyRs has helped to initially predict the localization of potential sites of interaction for modulators of channel activity, such as adenine nucleotides, calcium, calmodulin, and protein kinases.[1,2,64]

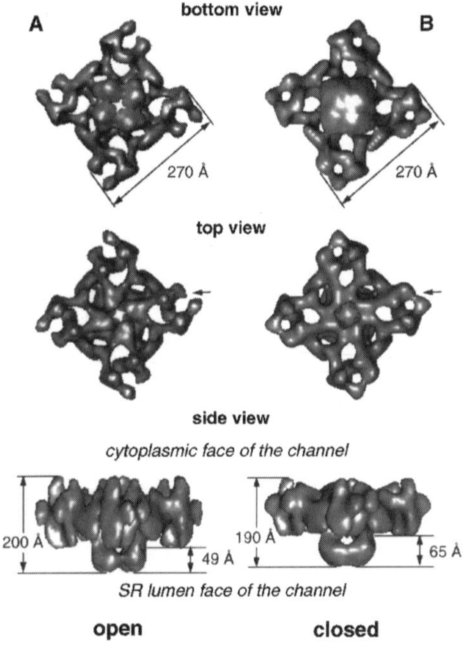

FIGURE 2. Bottom, top, and side views of the RyR1 in its open and closed states, as from electron cryomicroscopy studies (courtesy of Dr. Susan Hamilton).

Several sites for calmodulin-dependent protein kinase (CaM kinase) and cAMP-dependent protein kinase (PKA) are found in the RyR amino acid sequence, using the consensus RXX(S,T) and KRXX(S,T), respectively. Experimental evidence has shown that CaM kinase preferentially phosphorylates the S^{2809} in RyR2.[65,66] This has been demonstrated by direct sequencing of the phosphorylated tryptic fragment after *in vitro* phosphorylation of purified of RyR2 with CaM kinase.[66] PKA can also phosphorylate RyR2, but with a lower efficiency compared to CaM kinase.[67] Apparently, the same site also is phosphorylated in RyR1.[68,69] The relevance of phosphorylative events on RyR1 activity has been supported by patch-clamp experiments with sarcoplasmic reticulum preparations, where phosphatase treatment, or a peptide designed to specifically inhibit CaM kinase II, can reverse calcium-dependent inactivation of calcium release.[70]

Potential binding sites for calcium, ATP, and calmodulin were predicted in the region between residues 4253 and 4499 of the rabbit RyR1.[5,6] Antibodies that react with the region 4445-4586 reduce the open probability of the RyR1 channel.[55] $^{45}Ca^{2+}$ overlay of fusion proteins covering the length of RyR1 indicated the presence of a calcium binding site in the region between aa 4478-4512.[56] Antibodies against this sequence increase the open probability and the opening time of purified RyR1 incorporated in planar lipid bilayers — in line with a role of this site in calcium-dependent regulation of the channel. A second antibody against the peptide sequence PEPEPEPEPE (corresponding to amino acids residues 4489-4499) inhibits calcium- or caffeine-stimulated channel activities, but does not prevent ATP-induced activation of the channel.[57]

In addition to the above-described site in RyR1, other calcium binding sites have been predicted in all three RyR sequences at regions that resemble EF-hand motifs. Several of these sites also have been directly demonstrated to be able to bind Ca.[56-58,71] Calmodulin binding has been found on RyR amino acid sequences and, more recently, calmodulin binding at several of these predicted binding sites has been demonstrated.[71,72]

Out of several potential sites for ATP binding that are present on each RyR isoform, only one seems to be well conserved (amino acids 2237-2242 of mink RyR3, and corresponding amino acids of the rabbit RyR1 and RyR2 protein sequences). Using a photoactive derivative of ATP, it has been possible to identify the major ATP binding site in the RyR1 protein.[73] Photoaffinity labeling resulted in the covalent binding of labeled ATP to RyR1. Controlled trypsin digestion resulted in the accumulation of this compound in a fragment with a molecular weight of 76 kDa. Interestingly, with a similar technical approach (i.e., using an azido derivative of Ry, another agonist of calcium channel activity of RyRs) the site of binding of Ry on the RyR1 has been located on a 76-kDa fragment of RyR1 derived from the carboxyl-terminal part of the molecule.[74] The 76-kDa fragment, which probably contains some of the transmembrane domains of the RyR, would then contain binding sites for Ry, ATP, and calcium.

7. TRANSMEMBRANE ASSEMBLY

In the C-terminal of RyRs amino acid sequence up to 12 potential transmembrane hydrophobic sequences can be identified (TM′, TM″, and TM1-TM10). Based on structure prediction algorithms, two models have been proposed: in the first,[5] only the four larger and most hydrophobic sequences represent transmembrane domains (TM5, TM6, TM8, TM10); the second one suggests ten transmembrane domains (TM1-TM10).[6] How many of these domains really do span the membrane of the endoplasmic reticulum, or of the organelle where these channels are located, has not yet been experimentally proved. However, it is accepted that both the N- and the C-terminal part of these molecules are exposed in the cytosol, in analogy to experimental data obtained originally for the $InsP_3$-Rs and, more recently, also for RyR1.[75]

In a recent report by Susan Hamilton and co-workers,[76] it has been suggested that the region following Arg-4475 in RyR1 may still possess calcium channel activity. Although this evidence is rather indirect, it implies that only the transmembrane segments present in this region of the protein (i.e., all of the 4 TM domains of the first model, but only the last 6 of the second model) may participate to form the calcium pore of RyR. There are currently two models for the $InsP_3$-R transmembrane assembly, one with six and another with eight transmembrane domains.[77,78] It will be important to verify experimentally the real number of regions that cross the membrane in these two families of channels, and to compare the overall organization of the transmembrane domains and of the actual calcium pore they form. Given the conserved membrane topology of other ion channels, it can be expected that they are more similar in this respect than we can presently appreciate. In Figure 1, all 12 hydrophobic regions present in the RyR amino acid sequences have been reported. The four most hydrophobic domains are solid black and the less hydrophobic have been filled with two gradations of gray according to the relative hydrophobicity profiles of the remaining potential transmembrane domains, with the lighter ones being those with a minor probability of passing the membrane.

If the transmembrane domain assembly of RyRs is not clear, even less is known about the pore region of these channels. Searching for homology between RyRs and other known proteins, a low level of homology allows us to superimpose the segments 8 and 10 of RyRs with the S4/H4 and S5/H6 regions of the sodium channels and with the corresponding sequences of the dihydropyridine-sensitive calcium channels.[5,6] A similar analysis can be made for the $InsP_3$-Rs. It is interesting to note that segment 9 in RyRs would then correspond to the H5 region that in sodium and calcium channels has been shown to contribute to the wall of the actual pore of the channel.[79,80] On the basis of this homology, the equivalent region in the $InsP_3$-Rs has been proposed to be involved in forming the actual calcium pore, and thus has been referred to as "P region". Continuing with this analogy, Yamamoto-Hino et al.[81] pointed out that the aspartic acid following the glycine in the sequence RAGGI*GD* in the pore region of $InsP_3$, that is highly conserved in segment 9 of RyRs, may correspond to the acidic residues positioned in the "pore region" of voltage-dependent calcium channels, which have been shown by site-directed mutagenesis to direct the calcium preference of these channels.[82] Although the sequence homology in these regions is rather weak it should be possible to verify the potential importance of these residues in RyRs by site-specific mutagenesis.

8. CONCLUSION

Our knowledge of RyRs has much increased in these last few years, as both molecular biology and biochemistry have contributed to our understanding of these molecules. However, much remains to be done in the future, especially in order to better understand the structure/function relationship of these channels, and to dissect the sites that direct the specific properties of each isoform. Molecular biology approaches will contribute to these themes as well as in the development of new strategies to study these channels.

ACKNOWLEDGMENTS

I want to thank all present and past members of my lab, especially Giuseppe Giannini and Giovanna Marziali, who have actively contributed to the work discussed in this chapter, and Telethon, AIRC, CNR, and MURST for economic support.

REFERENCES

1. **Coronado, R., Morrissette, J., Sukhareva, M., and Vaugham, D. M.,** Structure and function of ryanodine receptors, *Am. J. Physiol.,* 266, C1485, 1994.
2. **Meissner, G.,** Ryanodine receptor/Ca^{2+} release channels and their regulation by endogenous effectors, *Annu. Rev. Physiol.,* 56, 485, 1994.
3. **Mikoshiba, K.,** Inositol 1,4,5-trisphosphate receptor, *Trends Pharmacol. Sci.,* 14, 86, 1993.
4. **Marks, A. R., Tempst, P., Hwang, K. S., Taubman, M. B., Inui, M., Chadwick, C., Fleischer, S., and Nadal-Ginard, B.,** Molecular cloning and characterization of the ryanodine receptor/junctional channel complex cDNA from skeletal muscle sarcoplasmic reticulum, *Proc. Natl. Acad. Sci. U.S.A.,* 86, 8683, 1989.
5. **Takeshima, H., Nishimura, S., Matsumoto, T., Ishida, H., Kangawa, K., Minamino, N., Matsuo, H., Ueda, M., Hanaoka, M., Hirose, T., and Numa, S.,** Primary structure and expression from complementary DNA of skeletal muscle ryanodine receptor, *Nature,* 439, 1989.
6. **Zorzato, F., Fujii, J., Otsu, K., Phillips, M., Green, N. M., Lai, F. A., Meissner, G., and MacLennan, D. H.,** Molecular cloning of cDNA encoding human and rabbit forms of the Ca^{2+} release channel (ryanodine receptor) of skeletal muscle sarcoplasmic reticulum, *J. Biol. Chem.,* 265, 2244, 1990.
7. **Fujii, J., Otsu, K., Zorzato, F., Leon, S. D., Khanna, V. K., Weiler, J. E., O'Brien, P. J., and MacLennan, D. H.,** Identification of a mutation in porcine ryanodine receptor associated with malignant hyperthermia, *Science,* 25, 448, 1991.
8. **Takeshima, H., Nishimura, S., Nishi, M., Ikeda, M., and Sugimoto, T.,** A brain-specific transcript from the 3′-terminal region of the skeletal muscle ryanodine receptor gene, *Fed. Eur. Biochem. Soc.,* 322, 105, 1993.
9. **Chen, S. R. W., Vaughan, D. M., Arey, J. A., Coronado, R., and MacLennan, D. H.,** Functional expression of cDNA encoding the Ca^{2+} release channel (ryanodine receptor) of rabbit skeletal muscle sarcoplasmic reticulum in COS-1 cells, *Biochemistry,* 32, 3743, 1993.
10. **Penner, R., Neher, E., Takeshima, H., Nishimura, S., and Numa, S.,** Functional expression of the calcium release channel from skeletal muscle ryanodine receptor cDNA, *FEBS Lett.,* 259, 217, 1989.
11. **Brillantes, A. B., Ondrias, K., Scott, A., Kobrinsky, E., Ondriasovà, E., Moschella, M. C., Jayaraman, T., Landers, M., Ehrlich, B. E., and Marks, A. R.,** Stabilization of calcium release channel (ryanodine receptor) function by FK-506-binding protein, *Cell,* 77, 513, 1994.
12. **Nakai, J., Imagawa, T., Hakamata, Y., Shigekawa, M., Takeshima, H., and Numa, S.,** Primary structure and functional expression from cDNA of the cardiac ryanodine receptor/calcium release channel, *FEBS Lett.,* 271, 169/177, 1990.
13. **Otsu, K., Willard, H. F., Khanna, V. K., Zorzato, F., Green, N. M., and MacLennan, D. H.,** Molecular cloning of cDNA encoding the Ca^{2+} release channel (ryanodine receptor) of rabbit cardiac muscle sarcoplasmic reticulum, *J. Biol. Chem.,* 265, 13472, 1990.

14. **Giannini, G., Clementi, E., Ceci, R., Marziali, G., and Sorrentino, V.,** Expression of a ryanodine receptor-Ca^{2+} channel that is regulated by TGF-β, *Science,* 257, 91, 1992.
15. **Hakamata, Y., Nakai, J., Takeshima, H., and Imoto, K.,** Primary structure and distribution of a novel ryanodine receptor/calcium release channel from rabbit brain, *FEBS Lett.,* 312, 229, 1992.
16. **Marziali, G., Rossi, D., Giannini, G., Charlesworth, A., and Sorrentino, V.,** cDNA cloning of mink RyR3 mRNA reveals multiple splicing sites, submitted.
17. **Giannini, G., Conti, A., Mammarella, S., Scrobogna, M., and Sorrentino, V.,** The ryanodine receptor/calcium channel genes are widely and differentially expressed in murine brain and peripheral tissues, *J. Cell Biol.,* 128, 893, 1995.
18. **McPherson, P. S. and Campbell, K. P.,** Characterization of the major brain form of the ryanodine receptor/Ca^{2+} release channel, *J. Biol. Chem.,* 268, 19785, 1993.
19. **MacLennan, D. H., Duff, C., Zorzato, F., Fujii, J., Phillips, M., Korneluk, R. G., Frodis, W., Britt, B. A., and Worton, R. G.,** Ryanodine receptor gene is candidate for predisposition to malignant hyperthermia, *Nature,* 343, 559, 1990.
20. **McCarthy, T., Healy, J. M. S., Heffron, J. J. A., Lehane, M., Deufel, T., Lehmann-Horn, F., Farral, M., and Johnson, K.,** Localization of the malignant hyperthermia susceptibility locus to human chromosome 19q12-13.2, *Nature,* 343, 562, 1990.
21. **Otsu, K., Fujii, J., Periasamy, M., Difilippantonio, M., Uppender, M., Ward, D. C., and MacLennan, D. H.,** Chromosome mapping of five human cardiac and skeletal muscle sarcoplasmic reticulum protein genes, *Genomics,* 17, 507, 1993.
22. **Sorrentino, V., Giannini, G., Malzac, P., and Mattei, M. G.,** Localization of a novel ryanodine receptor gene (RYR3) to human chromosome 15q14-q15 by *in situ* hybridization, *Genomics,* 18, 163, 1993.
23. **Cavanna, J. S., Greenfield, A. J., Johnson, K. J., Marks, A. R., Nadal-Ginard, B., and Brown, S. D. M.,** Establishment of the mouse chromosome 7 region with homology to the myotonic dystrophy region of human chromosome 19q, *Genomics,* 7, 12, 1990.
24. **Mattei, M. G., Giannini, G., Moscatelli, F., and Sorrentino, V.,** Chromosomal localization of murine ryanodine receptor genes RYR1, RYR2, and RYR3 by *in situ* hybridization, *Genomics,* 22, 202, 1994.
25. **Takeshima, H., Iino, M., Takekura, H., Nishi, M., Kuno, J., Minowa, O., Takano, H., and Noda, T.,** Excitation-contraction uncoupling and muscular degeneration in mice lacking functional skeletal muscle ryanodine-receptor gene, *Nature,* 369, 556, 1994.
26. **Catterall, W. A.,** Excitation-contraction coupling in vertebrate skeletal muscle: a tale of two calcium channels, *Cell,* 64, 871, 1991.
27. **Olivares, E. B., Tanksley, S. J., Airey, J. A., Beck, C. F., Ouyang, Y., Deerinck, T. J., Ellisman, M. H., and Sutko, J. L.,** Nonmammalian vertebrate skeletal muscles express two triad junctional foot protein isoforms, *Biophys. J.,* 59, 1153, 1991.
28. **Airey, J. A., Beck, C. F., Murakami, K., Tanksley, S. J., Deerinck, T. J., Ellisman, M. H., and Sutko, J. L.,** Identification and localization of two triad junctional foot protein isoforms in mature avian fast twitch skeletal muscle, *J. Biol. Chem.,* 265, 14187, 1990.
29. **Dutro, S. M., Arey, J. A., Beck, C. F., Sutko, J. L., and Trumble, W. R.,** Ryanodine receptor expression in embryonic avian cardiac muscle, *Dev. Biol.,* 155, 431, 1993.
30. **Airey, J. A., Baring, M. D., and Sutko, J. L.,** Ryanodine receptor protein is expressed during differentiation in the muscle cell lines BC_3H1 and C2C12, *Dev. Biol.,* 148, 365, 1991.
31. **Airey, J. A., Grinsell, M. M., Jones, L. R., Sutko, J. L., and Witcher, D.,** Three ryanodine receptor isoforms exist in avian striated muscles, *Biochemistry,* 32, 5739, 1993.
32. **Walton, P. D., Arey, J. A., Sutko, J. L., Beck, C. F., Mignery, G. A., Sudhof, T. C., Deerinck, T. J., and Ellisman, M. H.,** Ryanodine and inositol trisphosphate receptors coexist in avian cerbellar purkinje neurons, *J. Cell Biol.,* 113, 1145, 1991.
33. **Ellisman, H. S., Deerinck, T. J., Ouyang, Y., Beck, C. F., Tanksley, S. J., Walton, P. D., Airey, J. A., and Sutko, J. L.,** Identification and localization of ryanodine binding proteins in the avian central nervous system, *Neuron,* 5, 135, 1990.

34. **Sutko, J. L., Airey, J. A., Murakami, K., Takeda, M., Beck, C., Deerinck, T. J., and Ellisman, M. H.,** Foot protein isoforms are expressed at different times during embryonic chick skeletal muscle development, *J. Cell Biol.,* 113, 793, 1991.
35. **Airey, J. A., Baring, M. D., Beck, C. F., Chelliah, Y., Deerinck, T. J., Ellisman, M. H., Houenou, L. J., McKemy, D. D., Sutko, J. L., and Talvenheimo, J.,** Failure to make normal α ryanodine receptor is an early event associated with the crooked neck dwarf (cn) mutation in chicken, *Dev. Dynamics,* 197, 169, 1993.
36. **Airey, J. A., Deerinck, T. J., Ellisman, M. H., Houenou, L. J., Ivanenko, A., Kenyon, J. L., McKemy, D. D., and Sutko, J. L.,** Crooked neck dwarf (cn) mutant chicken skeletal muscle cells in low density primary cultures fail to express normal α ryanodine receptor and exhibit a partial mutant phenotype, *Dev. Dynamics,* 197, 189, 1993.
37. **Oyamada, H., Murayama, T., Takagi, T., Iino, M., Iwabe, N., Miyata, T., Ogawa, Y., and Endo, M.,** Primary structure and distribution of ryanodine-binding protein isoforms of the bullfrog skeletal muscle, *J. Biol. Chem.,* 269, 17206, 1994.
38. **Ottini, L., Marziali, G., Conti, A., Charlesworth, A., and Sorrentino, V.,** submitted.
39. **Takeshima, H., Nishi, M., Iwabe, N., Miyata, T., Hosoya, T., Masai, I., and Hotta, Y.,** Isolation and characterization of a gene for a ryanodine receptor/calcium release channel in *Drosophila melanogaster, FEBS Lett.,* 337, 81, 1994.
40. **Hasan, G. and Rosbash, M.,** Drosophila homologs of two mammalian intracellular Ca^{2+}-release channels: identification and expression patterns of the inositol 1,4,5-triphosphate and the ryanodine receptor genes, *Development,* 166, 967, 1992.
41. **Seok, J.-H., Xu, L., Kramarcy, N. R., Sealock, R., and Meissner, G.,** The 30 S lobster skeletal muscle Ca^{2+} release channel (ryanodine receptor) has functional properties distinct from the mammalian channel proteins, *J. Biol. Chem.,* 267, 15893, 1992.
42. **Kim, Y. K., Valdivia, H. H., Maryon, E. B., Anderson, P., and Coronado, R.,** High molecular weight proteins in the nematode *C. elegans* bind (^{3}H) ryanodine and form a large conductance channel, *Biophys. J.,* 63, 1379, 1992.
43. **Waterston, R., Martin, C., Craxton, M., Huynh, C., Coulson, A., Hillier, L., Durbin, R., Green, P., Shownkeen, R., Halloran, N., Metzstein, M., Hawkins, T., Wilson, R., Berks, M., Du, Z., Thomas, K., Thierry-Mieg, J., and Sulston, J.,** A survey of expressed genes in *Caenorhabditis elegans, Nature Genetics,* 1, 114, 1992.
44. **Berridge, M. J.,** Inositol trisphosphate and calcium signaling, *Nature,* 361, 315, 1993.
45. **Kasai, H. and Petersen, O. H.,** Spatial dynamics of second messengers: IP3 and cAMP as long-range and associative messengers, *TINS,* 17, 95, 1994.
46. **O'Brien, J., Meissner, G., and Block, B. A.,** The fastest contracting muscles of nonmammalian vertebrates express only one isoform of the ryanodine receptor, *Biophys. J.,* 65, 2418, 1993.
47. **Percival, A. L., Williams, A. J., Kenyon, J. L., Grinsell, M. M., Airey, J. A., and Sutko, J. L.,** Chicken skeletal muscle ryanodine receptor isoforms/ion channel properties, *Biophys. J.,* 67, 1834, 1994.
48. **Zhang, Y., Chen, H. S., Khanna, V. K., Leon, S. D., Phillips, M. S., Shappert, K., Britt, B. A., Brownell, A. K. W., and MacLennan, D. H.,** A mutation in the human ryanodine receptor gene associated with central core disease, *Nature Genetics,* 5, 46, 1993.
49. **Zorzato, F., Sacchetto, R., and Margreth, A.,** Identification of two ryanodine receptor transcript in neonatal, slow-, and fast-twitch rabbit skeletal muscles, *Biochem. Biophys. Res. Commun.,* 203, 1725, 1994.
50. **Giros, B., Martres, M. P., Pilon, C., Sokoloff, P., and Schwartz, J. C.,** Shorter variations of the D3 dopamine receptor produced through various pattern of alternative splicing, *Biochem. Biophys. Res. Commun.,* 176, 1584, 1991.
51. **Ferris, C. D., Huganir, R. L., Bredt, D. S., Cameron, A. M., and Snyder, S. H.,** Inositol trisphosphate receptor: phosphorylation by protein kinase C and calcium calmodulin-dependent protein kinase in reconstituted lipid vesicle, *Proc. Natl. Acad. Sci. U.S.A.,* 88, 2232, 1991.

52. **Danoff, S. K., Ferris, C. D., Donath, C., Fischer, G. A., Munemitsu, S., Ullrich, A., Snyder, S. H., and Ross, C. A.,** Inositol 1,4,5-triphosphate receptors: distinct neuronal and non-neuronal forms derived by alternative splicing differ in phosphorylation, *Proc. Natl. Acad. Sci. U.S.A.,* 88, 2951, 1991.
53. **Ferris, C. D. and Snyder, S. H.,** Inositol 1,4,5-trisphosphate-activated calcium channels, *Annu. Rev. Physiol.,* 54, 469, 1992.
54. **Sorrentino, V. and Volpe, P.,** Ryanodine Receptors: how many, where and why?, *TiPS,* 14, 98, 1993.
55. **Fill, M., Mejia-Alvarez, R., Zorzato, F., Volpe, P., and Stefani, E.,** Antibodies as probes for ligand gating of single sarcoplasmic reticulum Ca^{2+}-release channels, *Biochem. J.,* 273, 449, 1991.
56. **Chen, S. R. W., Zhang, L., and Lennan, D. M.,** Characterization of a Ca^{2+} binding and regulatory site in the Ca^{2+} release channel (ryanodine receptor) of rabbit skeletal muscle sarcoplasmic reticulum, *J. Biol. Chem.,* 267, 23318, 1992.
57. **Chen, S. R. W., Zhang, L., and MacLennan, D. H.,** Antibodies as probes for Ca^{2+} activation sites in the Ca^{2+} release channel (ryanodine receptor) of rabbit skeletal muscle sarcoplasmic reticulum, *J. Biol. Chem.,* 268, 13414, 1993.
58. **Treves, S., Chiozzi, P., and Zorzato, F.,** Identification of the domain by anti-(ryanodine receptor) antibodies which affect Ca^{+2}-induced Ca^{2+} release, *Biochem. J.,* 291, 757, 1993.
59. **Wagenknecht, T., Grassucci, R., Frank, J., Saito, A., Inui, M., and Fleischer, S.,** Three dimensional architecture of the calcium channel/foot structure of sarcoplasmic reticulum, *Nature (London),* 338, 167, 1989.
60. **Franzini-Amstrong, C.,** Studies of the triad. I. Structure of the junction in frog twitch fibers, *J. Cell. Biol.,* 47, 488, 1970.
61. **Ferguson, D. G., Schwartz, H., and Franzini-Amstrong, C.,** Subunit structure of junctional feet in triads of skeletal muscle. A freeze-drying, rotary-shadowing study, *J. Cell. Biol.,* 99, 1735, 1984.
62. **Franzini-Amstrong, C. and Jorgensen, A. O.,** Structure and development of E-C coupling units in skeletal muscle, *Annu. Rev. Physiol.,* 56, 509, 1994.
63. **Radermacher, M., Rao, V., Grassucci, R., Frank, J., Timerman, A. P., Fleischer, S., and Wagenknecht, T.,** Cryo-electron microscopy and three-dimensional reconstruction of the calcium release channel/ryanodine receptor from skeletal muscle, *J. Cell Biol.,* 127, 411, 1994.
64. **Ehrlich, B. E., Kaftan, E., Bezprovannaya, S., and Bezprozvanny, I.,** The pharmacology of intracellular Ca^{2+}-release channels, *TiPS,* 15, 145, 1994.
65. **Witcher, D., Kovacs, R., Schulman, H., Cefali, D., and Jones, L.,** Unique phosphorylation site on the cardiac ryanodine receptor regulates calcium channel activity, *J. Biol. Chem.,* 266, 11144, 1991.
66. **Witcher, D., Strifler, B., and Jones, L.,** Cardiac specific phosphorylation site for multifunctional Ca^{2+}/calmodulin-dependent protein kinase is conserved in the brain ryanodine receptor, *J. Biol. Chem.,* 267, 4963, 1992.
67. **Yoshida, A., Takahashi, M., Imagawa, T., Shigekawa, M., Takisawa, H., and Nakamura, T.,** Phosphorylation of ryanodine receptors in rat myocytes during beta-adrenergic stimulation., *J. Biochem. Tokyo,* 111, 186, 1992.
68. **Strand, M., Louis, C., and Mickleson, J.,** Phosphorylation of the porcine skeletal and cardiac muscle sarcoplasmic reticulum ryanodine receptor, *Biochim. Biophys. Acta,* 1175, 319, 1993.
69. **Suko, J., Maurer-Fogy, I., Plank, B., Bertel, O., Wyskovsky, W., Hohenegger, M., and Hellmann, G.,** Phosphorylation of serine 2843 in ryanodine receptor-calcium release channel of skeletal muscle by cAMP-, cGMP-, and CaM-dependent protein kinase, *Biochim. Biophys. Acta,* 1175, 193, 1993.
70. **Wang, J. and Best, P. M.,** Inactivation of the sarcoplasmic reticulum calcium channel by protein kinase, *Nature (London),* 359, 739, 1992.

71. **Chen, S. R. W. and MacLennan, D. H.,** Identification of calmodulin-, Ca^{2+}-, and ruthenium red-binding domains in the Ca^{2+} release channel (ryanodine receptor) of rabbit skeletal muscle sarcoplasmic reticulum, *J. Biol. Chem.,* 269, 22698, 1994.
72. **Menegazzi, P., Larini, F., Treves, S., Guerrini, R., Quadroni, M., and Zorzato, F.,** Identification and characterization of three calmodulin binding sites of the skeletal muscle ryanodine receptor, *Biochemistry,* 33, 9078, 1994.
73. **Zarka, A. and Shoshan-Barmatz, V.,** Characterization and photoaffinity labeling of the ATP binding site of the ryanodine receptor from skeletal muscle, *Eur. J. Biochem.,* 213, 147, 1993.
74. **Witcher, D. R., McPherson, P. S., Kahl, S. D., Lewis, T., Bentley, P., Mullinnix, M. J., Windass, J. D., and Campbell, K. P.,** Photoaffinity labeling of the ryanodine receptor: Ca^{2+} release channel with an azido derivative of ryanodine, *J. Biol. Chem.,* 269, 1376, 1994.
75. **Marty, I., Villaz, M., Arlaud, G., Bally, I., and Ronjat, M.,** Transmembrane orientation of the N-terminal and C-terminal ends of the ryanodine receptor in the sarcoplasmic reticulum of rabbit skeletal muscle, *Biochem. J.,* 298, 743, 1994.
76. **Callaway, C., Seryshev, A., Wang, J. P., Slavik, K. J., Needleman, D. H., Cantu, C., Wu, Y., Jayaraman, T., Marks, A. R., and Hamilton, S. L.,** Localization of the high and low affinity [3H]ryanodine binding sites on the skeletal muscle Ca^{2+} release channel, *J. Biol. Chem.,* 269, 15876, 1994.
77. **Kume, S., Muto, A., Aruga, J., Nakagawa, T., Michikawa, T., Furuichi, T., Nakade, S., Okano, H., and Mikoshiba, K.,** The Xenopus IP_3 receptor: structure, function, and localization in oocytes and eggs, *Cell,* 73, 555, 1993.
78. **Blondel, O., Takeda, J., Janssen, H., Seino, S., and Bell, G. I.,** Sequence and functional characterization of a third inositol trisphosphate receptor subtype, IP3R-3, expressed in pancreatic islets, kidney, gastrointestinal tract, and other tissues, *J. Biol. Chem.,* 268, 11356, 1993.
79. **Jan, L. Y. and Jan, Y. N.,** Superfamily of ion channels, *Nature,* 345, 672, 1990.
80. **Miller, C.,** Hunting for the pore of voltage-gated channels, *Curr. Biol.,* 2, 573, 1992.
81. **Yamamoto-Hino, M., Sugiyama, T., Hikinchi, K., Mattei, M. G., Hasegawa, K., Sekine, S., Sakurada, K., Miyawakis, A., Furuichi, T., Hasegawa, M., and Mikoshiba, K.,** Cloning and characterization of human type 2 and type 3 inositol 1,4,5-triphosphate receptors, *Receptors Channels,* 2, 9, 1994.
82. **Heinemann, S. H., Terlau, H., Stuhmer, W., Imoto, K., and Numa, S.,** Calcium channel characteristics conferred on the sodium channel by single mutations, *Nature,* 356, 441, 1992.
83. **Serysheva, I. I., Orlova, E. V., Chiu, W., Sherman, M. B., Hamilton, S. L., and von Heel, M.,** Electron cryomicroscopy and angular reconstruction used to visualize the skeletal muscle calcium release channel, *Nat. Struct. Biol.,* 2, 18, 1995.
84. **Furuichi, T., Furutama, D., Hakamata, Y., Nakai, J., Takeshima, H., and Mikoshiba, K.,** Multiple types of ryanodine receptor/Ca^{2+} release channels are differentially expressed in rabbit brain, *J. Neurosc.,* 14, 4794, 1994.

Chapter 7

EXPRESSION OF RYANODINE RECEPTORS

Alessandra Nori, Luisa Gorza, and Pompeo Volpe

CONTENTS

0-8493-8543-1/95/$0.00+$.50

1. INTRODUCTION

The present chapter reviews the expression of ryanodine receptors (RyRs)/Ca^{2+} release channels with emphasis on: (1) the differential distribution of distinct isoforms either in different tissues or in different cells of the same tissue, and (2) the physiological relevance of such a polymorphism to the regulation of intracellular Ca^{2+} homeostasis.

There are two classes of intracellular Ca^{2+} release channels, the RyRs and the receptors gated by the second messenger inositol-1,4,5-triphosphate (IP_3-Rs), each regulated by different agonists and activated by distinct pathways. Review is also made concerning the coexpression of each RyR isoform with IP_3-Rs and interaction of members of the two classes of Ca^{2+} release channels in the control of Ca^{2+} homeostasis.

Gene expression and localization of its product can be directly studied either at the level of RNA, using both DNA and RNA probes, or at the protein level, using monoclonal and polyclonal antibodies. Design of specific probes and sensitivity of the experimental procedures are crucial when more isoforms and low levels of expression are known or supposed. For the RyR class, these tools have become available very recently and have replaced the biochemical, pharmacological, and electrophysiological approaches used in the past to acquire indirect information on RyR expression. This section will discuss data obtained from Northern blot, *in situ* hybridization, RNA mapping for the RNA level, and Western blot, immunohistochemistry, and immunoelectron microscopy for the protein level.

The discovery, isolation, characterization, and cloning of the skeletal- and cardiac-muscle ryanodine receptors (RyR1 and RyR2, respectively) and, more recently, cloning of the "brain" isoform (RyR3), have made available specific probes for the three distinct isoforms. Data obtained in several different tissues, from both mammals and nonmammals, indicate that RyR isoforms are expressed either alone or in combination: (1) with different ratios during development, (2) in some, but not all, tissues, (3) in specific sets of cells of given tissues, and (4) in distinct subcellular compartments.

The large number of experimental systems employed thus far allows only a fragmentary analysis of published data and makes it difficult, at present, to understand the biological relevance of coexpression of multiple isoforms. It is plausible to suppose, however, that each RyR isoform has distinct functional characteristics, and may even be regulated by slightly different mechanisms. Therefore, RyR polymorphism could conceivably afford a more plastic modulation of cell responses to precise stimuli compared to that exerted by an invariant RyR form. Unraveling the expression pattern of different RyR isoforms and understanding their functional characteristics may help to understand the role(s) of this class of intracellular Ca^{2+} channels in cell physiology.

2. TISSUE AND CELL EXPRESSION OF RyRS

2.1. MAMMALS

The distribution of RyRs mRNA in different tissues and/or regions and the approximate expression levels are summarized in Table 1; subregional localization and, when possible, cell distribution are also indicated.

2.1.1. Skeletal Muscle

The skeletal muscle ryanodine receptor (RyR1) represents the main form of Ca^{2+} release channel localized in the junctional area of the sarcoplasmic reticulum (SR). RNA expression has been studied in fast- and slow-twitch skeletal muscles[1-3] and its precise localization in the junctional part of SR terminal cisternae has been investigated in depth and thoroughly at the electron microscopy level.[4,5]

A variant form of the RyR1 transcript, due to the alternative splicing of a five-amino acid exon, has been identified in rabbit and man.[6,7] These two transcripts, differing for the Ala-Gly-Asp-Ala-Gln sequence, are coexpressed both in adult muscles (fast- and slow-twitch muscles) and in immature muscles.[8]

Recently, by virtue of a sensitive method (RNAse protection) and by means of specific nucleotide probes, a very low level of RyR3 mRNA has been detected in total RNA derived from skeletal muscle of both mink[9] and mouse.[53] As of now, the role of RyR3 in skeletal muscle is completely obscure.

2.1.2. Smooth Muscles

In smooth muscles, redistribution of Ca^{2+} from intracellular Ca^{2+} stores is mainly mediated by IP_3-gated Ca^{2+} channels. Nevertheless, multiple experimental approaches, such as [^{3}H]-ryanodine binding, partial biochemical purification, and channel reincorporation in planar bilayers, have conclusively shown RyRs to be present in smooth muscles.

Studies on the expression of distinct isoforms by Northern blot and RNAse protection of mRNA have shown that RyR3 is expressed in organs containing smooth muscle: aorta, esophagus, taenia coli, urinary bladder, uterus and ureter of the rabbit,[10] and the stomach, spleen, lung, kidney, ileum, and jejunum of the mink and mouse.[9,11]

Blot hybridization with cDNA probes specific for RyR1 and RyR2 have revealed a very weak signal in aorta and stomach mRNAs, respectively, whereas in liver, kidney, and utero mRNAs no signal was obtained with either probe.[2,3,12] The high sensitivity of the RNAse protection assay, however, has allowed us to ascertain that very low levels of RyR1 and RyR2 are expressed, for example, in kidney, stomach, gut, and thymus.[11]

The interpretation of these results, however, is limited by the source of RNA, which in most cases is derived from the whole organ. For instance, vascular endothelium, one of the "contaminants" of smooth muscles in these preparations, has been reported to express RyR2, at least in the guinea pig.[13]

TABLE 1
Expression of RyRs in Mammals

Tissue	Methods	RyR1	RyR2	RyR3	Species	Ref.
Skeletal muscle	Cloning, RNA blot	+	–	–	Rabbit, mink, mouse	1, 2, 3
	RNAse protection	+++	–	+	Mink, mouse	9, 11
Heart	RNA blot, cloning	+	+	–	Rabbit	12
	RNAse protection, *in situ* hybridization	–	++	+	Mink, mouse,	9, This chapter
Heart ventricle	Immunoblot	–	+	–	Dog, rabbit, guinea pig, rat, hamster	13, 52
Endocardial endothelium	Immunohistochemistry	+	+	ND	Guinea pig	13
	In situ hybridization	–	–	+	Mouse	This chapter
Vascular endothelium	Immunoblot, immunohistochemistry	+?	+?	–	Guinea pig	13
	In situ hybridization	–	+	–	Mouse	This chapter
Coronary vessels	*In situ* hybridization	–	–	+	Mouse	This chapter
Epithelial cells (lung)	Cloning	ND	ND	+	Mink	9
Stomach	RNA blot, RNAse protection	–/+	+	+	Rabbit, mink, mouse	9, 10, 11, 12
Aorta	RNA blot, cloning	+	–	+	Rabbit	10, 3
Esophagus	RNA blot	ND	ND	+	Rabbit	10
Taenia coli	RNA blot	ND	ND	+	Rabbit	10
Urinary bladder	RNA blot	ND	ND	+	Rabbit	10
Ureter	RNA blot	ND	ND	+	Rabbit	10
Uterus	RNA blot	–	–	+	Rabbit	10, 3
Spleen	RNAse protection, RNA blot	–	–	+	Mink, mouse	9, 11
Kidney	RNAse protection, RNA blot	–/+	–/+	+	Rabbit, mink, mouse	3, 9, 11
Lung	RNAse protection	–	–	+	Rabbit, mink, mouse	3, 9,11
Ileum	RNAse protection	–	–	+	Mink	9
Jejunum	RNAse protection	–	–	+	Mink, mouse	9, 11
	RNA blot	–	+?	+?	Mink	9
Cerebellum	Immunoblot, immunohistochemistry	+	–	–	Mouse	22
Cerebellum	RNA blot, *in situ* hybridization	+	++	–	Rabbit	20
Cerebellum	Imunohistochemistry	–	–	+	Rat	21, 39
Corpus striatum	RmNA blot	–	+	++	Rabbit	20
Hippocampus	Immunoblot, immunohistochemistry	–	+	–	Cow, mouse, rat	21, 22, 39
Hippocampus	RNA blot	–	–	+	Rabbit	10
Hippocampus	*In situ* hybridization	+	++	+	Rabbit, rat	20, 21

TABLE 1 (continued)
Expression of RyRs in Mammals

Tissue	Methods	RyR1	RyR2	RyR3	Species	Ref.
Thalamus	*In situ* hybridization, RNA blot	–/+	++	+++	Rabbit	20
Thalamus	Immunohistochemistry	–	+	–	Mouse, rat	39, 22
Cerebral cortex	*In situ* hybridization	++	+++	+	Rabbit, rat	20, 21
Cerebral cortex	RNA blot	–	+	–	Rabbit	20
Cerebral cortex	Immunohistochemistry	–	+	–	Rat, mouse	39, 22
Olfactory bulb	*In situ* hybridization	+	+++	+	Rabbit, rat	20, 21
Olfactory bulb	RNA blot	+	+	+	Rabbit	20
Olfactory bulb	Immunohistochemistry	–	+	–	Mouse, rat	39, 22
Brain stem	*In situ* hybridization	–/+	++	+	Rabbit, rat	20, 21
Brain stem	Immunohistochemistry	–	+	–	Mouse	22
Hypothalamus	*In situ* hybridization	+	+	+	Rabbit, rat	20, 21
Hypothalamus	Immunohistochemistry	–	+	–	Rat, mouse	39, 22
Amygdala	*In situ* hybridization	+	+	+	Rabbit, rat	20, 21
Amygdala	Immunohistochemistry	–	+	–	Rat, mouse	39, 22

Note: For *in situ* hybridization and Northern blot +++ high, ++ medium, + low, +/– slightly above the background expression. ND, not determined; ?, nonspecific for the isoforms.

It is also obvious that the higher the sensitivity of the method, the lower the possibility of identifying the cell origin of a specific mRNA. Thus, cell localization of RNA by *in situ* hybridization becomes essential in the interpretation. *In situ* hybridization has provided evidence that RyR3 is present only in uterus muscle wall and vas deferens, whereas spleen and spleen sinusoids seem to express RyR1 and RyR3, respectively[11] (see also next section). Overall, published data allow the tentative conclusion that RyR3 is the main isoform expressed in some smooth muscle-containing organs.

2.1.3. Cardiac Muscle

2.1.3.1. Heterogeneity of RyR Expression in Heart Cells

Using immunological and nucleic acid probes that distinguished between RyR1 and RyR2, Lesh et al.[13] described the expression of a type 2-like RyR in vascular and endocardial endothelium of the guinea pig heart. The possibility that a RyR other than RyR2 could be expressed in the mammalian heart was already suggested by the evidence of RyR3 mRNA accumulation by RNAse protection assay.[9] However, the lack of specific probes in the former study and the absence of information concerning the cell type expressing the RyR3 mRNA in the heart, prompted us to study the distribution of the distinct RyRs at the cellular level in order to determine whether they show a complementary or an overlapping distribution in different heart cells.[14] Essential to this approach was the use of type-specific nucleic acid probes, due to the high similarity existing between RyR3 and RyR2 mRNAs.[10,11]

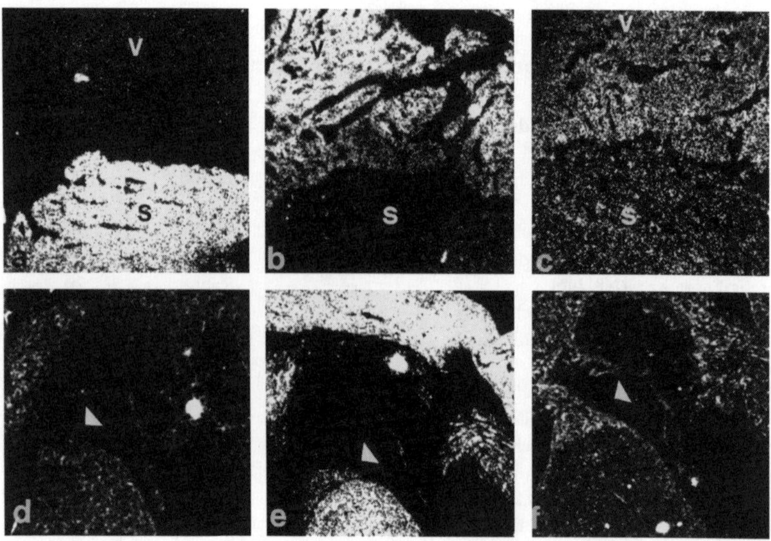

FIGURE 1. RyR mRNA expression in cardiac and skeletal muscles. Dark-field micrographs of cryosections of rat heart and skeletal muscle hybridized with antisense cRNA probes specific for RyR1 mRNA (**a,d**), RyR2 mRNA (**b,e**), and RyR3 mRNA (**c,f**). Skeletal muscle fibers (S) show strong signals for RyR1 mRNA, apparently no signal for RyR2 mRNA, and very weak signals for RyR3 mRNA; whereas cardiac myocytes (V) hybridize strongly with RyR2 probe, weakly with RyR3 probe, and do not show detectable hybridization with RyR1 probe. In **d** to **f** arrowhead indicates conduction myocytes of the atrioventricular bundle. Note that conduction myocytes display stronger hybridization signals for RyR3 and weaker signals for RyR2 than working myocytes.

In situ hybridization analysis with mouse cRNA probes specific for each RyR isoform showed that RyR1 mRNA is not expressed in any cardiac cell type (Figure 1a and d). Strong hybridization signals for RyR2 mRNA could be detected in working myocytes of the atrial and ventricular myocardium. In contrast, very low signals were detected in a subset of myocytes that belong to the heart conduction system (Figure 1b and d). The RyR3 probe showed weak hybridization signals in working myocytes and stronger signals in conduction myocytes (Figure 1c and e).

The RyR immunoreactivity in corbular SR of sheep conduction myocytes[15] could thus be interpreted as largely due to the expression of RyR3.

Accumulation of RyR3 mRNA, but not of RyR2 mRNA, was detected in the endocardial endothelium; in contrast, vascular endothelium showed weak hybridization signals for RyR2 mRNA (Figure 2a and Reference 13). Smooth muscle cells of intracardiac coronary vessels apparently did not show accumulation of RyR2 mRNA, whereas they hybridized strongly with RyR3 (Figure 2b) and, similarly to conduction myocytes, accumulated IP_3-R mRNA (Figure 2c; and Section 3).

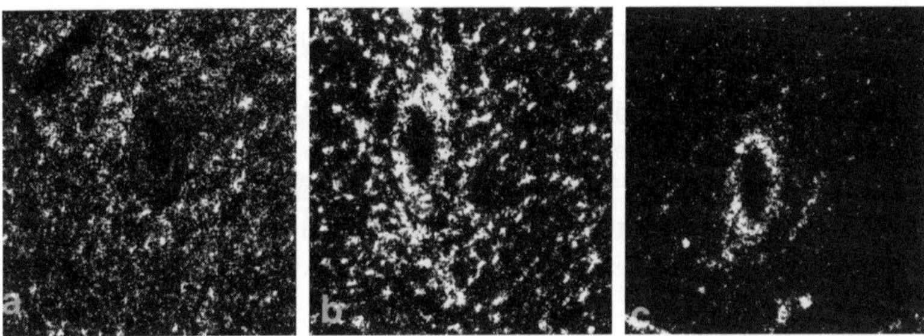

FIGURE 2. Intracellular Ca^{2+} release channel mRNAs in intracardiac coronary vessels. Rat heart cryosections are hybridized with probes specific for RyR2 (**a**) and RyR3 mRNAs (**b**), and for IP_3-R type1 mRNA (**c**). Smooth muscle cells of intracardiac coronary vessels show strong signals for RyR3 and for IP_3-R type 1 mRNAs, whereas they do not apparently hybridize with the RyR2 probe. In contrast, a weak hybridization signal for RyR2 mRNA is observed on vascular endothelial cells (**a**).

2.1.3.2. RyR Expression During Heart Development

Dramatic changes in excitation-contraction coupling are known to occur in the heart during the postnatal period. Increasing responsiveness to ryanodine and a lowering of the threshold of free Ca^{2+} for Ca^{2+}-induced-Ca^{2+} release (CICR) from the SR is accounted for by the increased amount of RyR2 mRNA and protein and by the development of T tubules, where both ryanodine and 1,4-dihydropyridine (DHP) receptors become concentrated.[16,17]

Expression of RyRs has been considered to play a marginal role during prenatal development. Nevertheless, a recent study showed that functional RyRs may be detected in the chick heart as early as day 4 of embryonic development.[18] Interestingly, exposure to ryanodine in early embryonic hearts affected mostly the chronotropism, whereas in late embryonic hearts ryanodine exerted only a negative inotropic effect. In preliminary experiments on embryonic rat hearts, precocious expression of RyR2 mRNA and delayed expression of RyR3 mRNA have been observed.[54] Whether RyR isoform transition is involved in modulation of chronotropic and inotropic effects awaits further experimental tests.

2.1.4. Brain

The brain represents an extremely complex example of the expression of RyR isoforms in distinct districts of the same organ. At least three genes are expressed in mammalian brain, as judged by Northern blot of RNA prepared from different regions.

Hakamata et al.,[10] who first cloned the RyR3 isoform in the rabbit, showed that RyR3 is abundantly expressed in restricted areas of the brain — hippocampus, corpus striatus, and thalamus — whereas RyR2 is expressed in other areas (see Table 1), in particular, the cerebellum. A truncated form of RyR1 transcript has been identified in brain total mRNA by Takeshima et al.[19] This cDNA encoding for the carboxyl-terminal 656 amino acid transmembrane domain of RyR1 could derive from an alternative transcription initiation codon (Met^{4382}).

Recently, *in situ* hybridization studies of rabbit brain with cRNA probes specific for RyR1, RyR2, and RyR3[20] have confirmed a heterogeneous distribution of neuronal RyRs. In the hippocampal formation, RYR2 RNA is highly expressed only in the dentate gyrus, in the granular cell layer and, at a low level, together with RyR1 and RyR3, in the pyramidal cells of the CA1-2-3-4 areas; the CA1 area seems particularly enriched in RyR3 RNA (as observed by Giannini et al.[11] in the mouse). High levels of RyR2 and low levels of RyR1 are expressed in cerebellum. Both isoforms are expressed in Purkinje cells, whereas only RyR2 is expressed in the granular layer and in some neurons of the molecular layer; RyR3 seems occasionally detectable in single Purkinje and granular cells. In cerebral cortex, RyR2 is the predominant isoform with a peculiar "laminar"-shaped distribution.

In situ hybridization also has been used in sagittal sections of rat brain with a RyR2 antisense RNA probe;[21] the cardiac isoform is distributed throughout the brain and is particularly enriched in the dentate gyrus of the hippocampal formation, as reported in the rabbit.

At the protein level, immunological studies in the mouse brain were performed with antibodies raised against synthetic peptides specific for RyR2 and RyR1+RyR2.[22,23] These data confirmed expression of RyR2 in many cerebrum districts, with stronger reactivity in the cerebral cortex and hippocampus; in hippocampus, pyramidal cells of the CA2-3 regions were stained more than CA1; in dentate gyrus, granule cells and some interneurones were labeled; in the olfactory bulb, a strong labeling of the mitral cells was reported.

RyR1 appeared to be expressed only in cerebellum. The antiserum recognizing both skeletal and cardiac isoforms, in fact, had a strong reactivity in Purkinje and Golgi cells in cerebellum, whereas other cells, such as stellate, basket, and granular cells, were weakly labeled. In these positive neurons the somata were mainly stained, and only in some Purkinje cells were the dendrites stained. Lai et al.,[21] using a monoclonal antibody specific for RyR2, showed that immunoreactive regions in rat forebrain, as in the mouse, were the hippocampus, cortex, caudate-putamen, septum, amygdala, thalamus, hypothalamus, cerebellum, and the region corresponding to the mossy fibers; labeling was observed in dendrite layers as well as in the somata.

In the work by Lai et al.,[21] Kuwajima et al.,[22] and Nakanishi et al.,[23] the antibody specificity did not distinguish RyR1 and RyR2 from RyR3; the lack of specificity can account for discrepancies observed with respect to studies of mRNA expression.

2.2. OTHER VERTEBRATES

2.2.1. Muscles

RyR receptors have been identified in other species of vertebrates: birds, amphibians, and fishes. Two RyR isoforms have been identified in skeletal muscles, and a third isoform was expressed in avian cardiac muscle.[24] The skeletal isoforms are called α and ß and the relation with the mammalian isoforms is still debated: very recent evidence indicates that αRyR is homologous to mammalian RyR1 and ßRyR is most homologous to mammalian RyR3.[25] Expression of α/ß RyR isoforms has been studied in muscle tissues mainly by immunological techniques. It has been shown by cloning of cDNA from bullfrog skeletal muscle that the α isoform is more than 85% homologous to that of the rabbit RyR1.[26]

An interesting point emerges from the comparison of the isoform pattern expression in specialized muscles of different species. Birds, fishes, and amphibians express similar amounts of both isoforms in skeletal muscles,[27,28] but two reptilian orders express only the α skeletal-like type.[29] The αRyR isoform is expressed in very fast-contracting muscles (extraocular muscles, swimbladder muscle) of fishes and birds. Thus, it would appear that expression of a single isoform takes place in very specialized muscles, i.e., as if it were necessary to maximize a given function, whereas two isoforms are expressed in muscles that rely upon a more modulative system. For example, the very fast contraction may require a strong coupling between excitation and contraction and this is accomplished by the skeletal-like α isoform; on the other hand, for the body musculature of ectothermic animals, the presence of similar amounts of channels with different physiological properties adaptable to a wide range of different temperatures, could be an advantage.

2.2.2. Cerebellum

Chick cerebellum has been investigated for the presence of RyRs by immunofluorescence and immunoelectron microscopy.[30,31] Antibodies used in these studies were reactive with both the α and ß skeletal muscle isoforms. It has been shown that Purkinje cells do express RyR proteins in the soma and dendritic shafts, whereas they seem absent from the dendritic spines, in which, on the contrary, IP_3-R is observed.

3. COEXPRESSION OF RyR AND IP_3-R

Coexpression of individual forms of both classes of intracellular Ca^{2+} release channels constitutes an intriguing yet unresolved aspect of molecular physiology. A still preliminary review is provided here with the aim of fostering a discussion on the physiological meaning, if any, of a given pattern of RyR and IP_3-R coexpression.

3.1. MOLECULAR HETEROGENEITY OF IP_3-R

Two levels of "molecular variability" in IP_3-R channels have been found by molecular cloning in mouse:

1. Several different genes encode for IP_3-sensitive Ca^{2+} channels: IP_3-RI (homologous in the rat to IP_3-R1); the most ubiquitous and highly expressed gene, IP_3-RII, IP_3-RIII (homologous in the rat to IP_3-R2), and IP_3-RIV,[32-34]
2. Multiple types of IP_3-Rs derive from two main alternative splicing sites (SI and SII) of IP_3-RI, and are referred to as IP_3-RSI-, IP_3-RSI+, IP_3-RSII ABC- (short), IP_3-RSII+ (long), IP_3-RSII B-, and IP_3-RSII BC-.[35]

The splicing events at the SI site lead to the presence or absence of a 15-amino acid stretch near the IP_3 binding domain which could be involved in modulation of IP_3 binding affinity to the receptor. The SII splice site is located between two potential PKA phosphorylation sequences and could be involved in alternative phosphorylation states of the channel. Interestingly, it has been shown that all these variant forms are expressed in a tissue-specific manner.

IP_3-RI is expressed at higher level than other IP_3-R genes; tissue distribution of IP_3-RI and of its alternatively spliced forms has been studied by Northern blot and PCR amplification of mRNA deriving from different tissues.[36-38] It has been reported that the long form is predominantly expressed in the nervous system (cerebellum, cerebral cortex, olfactory bulb, and trigeminal ganglion), whereas the short form is mainly expressed in other tissues. *In situ* hybridization and immunocytochemistry studies in the central nervous system by Shell et al.,[36] Furuichi et al.,[37] Nakagawa et al.,[38] and Sharp et al.[39] are reviewed below *vis a vis* with those concerning RyRs.

Expression levels of IP_3-RII and IP_3-RIV genes are very low; their tissue specificity has been studied by PCR and RNAse protection assay by Ross et al.[34] They are both expressed in brain (particularly cerebellum), spinal cord, testis, lung, placenta, liver, and kidney. Moreover, the same authors show that IP_3-RIII is widely distributed, with the highest levels in cerebellum Purkinje cells and in the smooth muscle layer of the gastrointestinal tract.

3.2. CONTROL OF Ca^{2+} RELEASE

Intracellular Ca^{2+} homeostasis is regulated by multiple, complex, and interacting mechanisms such as Ca^{2+} influx from the extracellular medium and uptake, storage, and release of Ca^{2+} from rapidly and slowly exchanging intracellular Ca^{2+} stores.[40]

Release of Ca^{2+} from rapidly exchanging Ca^{2+} stores via RyRs and IP_3-Rs is controlled by several distinct pathways, which are in some cases common to both classes of channels and among the isoforms, and specific in some other cases. Examples of regulation based on either specific localization or spliced and isoform expression of both channels are reported below.

Release of Ca^{2+} by IP_3-R is triggered by IP_3 that acts by increasing the open probability of the channel; the direct interaction between IP_3 and the channel is modulated *in vitro* by low "subphysiological" concentrations of ATP (that stimulates channel gating); this mechanism *in vivo* could be associated to a peculiar localization of the channel, for example, in cytoplasmic microdomains in the proximity of the Ca^{2+}-ATPase pump where ATP concentration can drop under the mean "physiological" levels. Another mechanism of channel regulation, the cAMP-dependent phosphorylation by PKA, modulates IP_3-R activity in a splice form-specific manner: the long form (expressed mainly in brain) is inhibited and the short form (prevalent in peripheral tissues but present also in brain) is activated by phosphorylation.

As to RyRs, a number of mechanisms can be mentioned: RyR1 in skeletal muscle junctional SR is coupled to the L-type voltage-gated Ca^{2+} channel dihydroprydine (DHP) receptor located in the T tubule. The DHP-R may act as a voltage sensor, allowing direct activation of RyR1 by membrane voltage (for details see next chapter). When RyR1 is dissociated from DHP-Rs *in vitro*, it behaves in a completely different way, being insensitive to voltage and sensitive to changes of [Ca^{2+}], like RyR2. Phosphorylation also plays a regulatory role in channel function, as recently reported for RyR1 and 2.[41]

cADP ribose is the nucleotide derivative that could play a physiological regulatory role on the RyRs (see Section 2). Its action seems isoform-specific since it has been shown to activate only RyR2.[42]

Ca^{2+} itself plays a key role in modulating Ca^{2+} release by both channels. Two Ca^{2+} binding sites are exposed on the cytoplasmic side of IP_3-R. A "bell-shaped" dependence on Ca^{2+} has been shown for IP_3-induced responses, with maximal activity around 250 n*M*, whereas a desensitization process occurs at higher concentrations.[43,44] On the lumenal side, the "filling status" of the store can be "sensed" by the channel. Modulation of channel activity from the lumenal side appears to rely upon interaction with yet another protein, since no Ca^{2+} binding sites are present on the receptor.

RyRs activity is highly modulated by Ca^{2+}. A deeply investigated phenomenon: the so-called Ca^{2+}-induced Ca^{2+} release (CICR) where rapid and considerable increases of cytoplasmic Ca^{2+} concentration induce activation of RyR2 and of isolated RyR1, and where high Ca^{2+} concentrations (in the millimolar range) inhibit, instead, all RyR isoforms.

In vivo Ca^{2+} release regulatory mechanisms involving RyR3 are still under investigation. Since ßRyR is structurally more similar to RyR3 than to RyR1,[26] ßRyR functional properties could be extended to RyR3. A comparison among the ion channel properties of αRyR and ßRyR has recently been performed using incorporation of purified channels in lipid bilayers.[25] These authors find a strong correspondence between αRyR and RyR1, and some differences in channel behaviors between ßRyR and αRyR: ßRyR channel openings are significantly longer than those of αRyR, and lower Ca^{2+} concentrations are needed to activate ßRyR (by decreasing the mean closed time); the gating behavior is more sensitive to Ca^{2+} concentration. However, the physiological relevance of such differences remains to be determined, particularly in relation to coexpression of two RyR isoforms in the same cell.

3.3. EXAMPLES OF COEXPRESSION OF LIGAND-GATED INTRACELLULAR Ca^{2+} CHANNELS AND IMPLICATIONS FOR CONTROL OF Ca^{2+} RELEASE

3.3.1. Immunocytochemistry

Sharp et al.[39] used an antibody specific for the rat RyR2 gene product and a polyclonal anti-IP_3-R antibody to compare RyR expression and IP_3-R expression in confocal and electron microscopy. RyR2 and IP_3-R were coexpressed in Purkinje cells and their proximal dendritic trees, whereas distal dendrites were positive only for IP_3-R; deep nuclei of cerebellum expressed RyR2 in cell bodies and IP_3-R in nerve terminals surrounding the cell bodies. Dentate gyrus and the hippocampal CA3 region were found to be enriched with RyR2 and weakly labeled for IP_3-R, whereas the CA1 region showed an inverse pattern. The motor cortex and posterior parietal cortex displayed a differential localization of RyR2 and IP_3-R: the first was evident in thin, long dendritic processes and the second was most prominent in cell bodies and proximal dendrites.

Electron microscopy analysis was performed on the hippocampal formation: RyR2 was present within dendritic spines, associated with saccules near the plasma membrane, whereas IP_3-R was predominant in the dendritic shafts.

3.3.2. *In Situ* Hybridization

3.3.2.1. Cerebellum

The Purkinje cells have been extensively used as models for studying Ca^{2+} homeostasis in neurons in that they express high levels of Ca^{2+} store components. In these cells, besides Ca^{2+} storage proteins[40] and the Ca^{2+} ATPase pump, both RyR-R and IP_3-R Ca^{2+} channels are highly expressed (see above). All the different splicing forms of IP_3-RI, except the short, are expressed in cell bodies and proximal dendrites; in dendrites and dendritic spines the long form is predominant. RyR1 and RyR2 are not expressed in dendritic spines.

In Purkinje cells, both IP_3-induced Ca^{2+} release and CICR occur upon application of appropriate stimuli. Production of IP_3 is stimulated by glutamate through the metabotropic glutamate receptors mGLUR1 and mGLUR5, which are coupled to plasma membrane phospholipase C in dendritic spines. This pathway could be involved in synaptic plasticity with mechanisms such as long-term depression (LTD, for a review see Reference 45). It has recently been shown that CICR can be induced in Purkinje cells by frequent pulses of membrane depolarization;[46] ruthenium red and caffeine affect CICR, thus implying that it depends on RyRs. It is plausible that the combination of IP_3-induced Ca^{2+} release and CICR can account for the long-lasting Ca^{2+} signal required for triggering LTD.

3.3.2.2. Hippocampus

The hippocampal formation expresses IP_3-RI and the three RyRs in all subregions with differences in the ratio among the distinct forms of both channels: dentate gyrus and CA3 region express high levels of RyR2 and low levels of IP_3-RI (short, long, and SI- forms); the CA1 region expresses high levels of IP_3-RI and is enriched in RyR3. At the subcellular level, pyramidal cells of the CA1 region contain IP_3-RI spliced forms B-, SI+, and long in the perinuclear area. The long form is present in proximal dendrites, but is absent in dendritic spines where RyR3 seems to be expressed.

The heterogeneous distribution of the Ca^{2+} channels in different hippocampal regions resembles the functional segregation of Ca^{2+} pools:[47] IP_3-sensitive Ca^{2+} pools are prevalent in the CA1 region, whereas caffeine-sensitive Ca^{2+} pools are enriched in CA3 and dentate gyrus. Furthermore, Verma et al.[47] have shown that hippocampal microsomes display both CICR and IP_3-induced Ca^{2+} release.

The role of Ca^{2+} release in long-term potentiation (LTP) of synaptic transmission in the hippocampus has been reviewed by Bliss and Collingridge.[48] The best understood form of LTP is induced by activation of the NMDA channels; the transient Ca^{2+} flux which permeates NMDA channels is subsequently amplified by Ca^{2+} release from intracellular stores. A parallel pathway which may be important for the induction of LTP is provided by the activation of mGLURs, IP_3 production, and IP_3-induced Ca^{2+} release.

3.3.2.3. *Cerebral Cortex*

Expression of IP_3-RI, particularly in the II and IV pyramidal cell layers, has been shown by *in situ* hybridization and immunocytochemistry. The spliced forms have not yet been identified. At the subcellular level it is expressed in the perinuclear cytoplasm, axons, and distal dendrites. Among the RyRs, only RyR2 is expressed with a similar (laminar-shaped) pattern, whereas RyR1 and RyR3, where detectable, are randomly distributed.

3.3.2.4. *Smooth Muscle*

IP_3-RI and RyR3 are clearly coexpressed in smooth muscle cells of the intracardiac coronary vessels (Figure 2 and Reference 14). Smooth muscle RyRs [51] have been reported to require higher Ca^{2+} concentrations for activation upon incorporation in lipid bilayers. Functional properties of such RyRs, probably but not surely RyR3, would be at variance with those of βRyRs and should, however, be considered in the context of RyR and IP_3-R coexpression, i.e., activation of IP_3-R as a prerequisite of RyR openings.

3.3.2.5. *Heart*

In addition to the preferential accumulation of RyR3, conduction myocytes also are enriched in IP_3-sensitive Ca^{2+} release channels.[49] The functional significance of RyR3 concentration in conduction myocytes, as in other heart cell types, remains to be determined. Studies performed on skinned dog myocytes have shown that conduction myocytes require a higher free Ca^{2+} concentration to induce Ca^{2+} release.[50] Thus, higher Ca^{2+} concentrations may be necessary to induce activation of RyR3[51] as it occurs after subliminal activation of IP_3-sensitive Ca^{2+} stores.[9]

4. CONCLUSIONS

Based on the reported examples, some general but still tentative rules can be envisioned:

1. RyR2 is always detected where CICR is observed, regardless of the presence of other RYR isoforms and of IP_3-Rs.
2. The ratio between RyR and IP_3-R isoforms is often quite different in cells of the same tissue; whether this reflects functional differences or subcellular compartmentalization awaits further experimental test.
3. RyR3 and the IP_3-RI long-spliced form appear to be coexpressed; evidence to this effect has been gathered in different cells and using distinct experimental approaches.

Given the established pattern of coexpression, RyR3 and the IP_3-RI long form may be the best-suited forms in cross-talk between separate Ca^{2+} pools. The nature of the interplay between IP_3-R and RyR remains to be fully investigated and requires studies of single cells expressing only two specific RyR and IP_3-R isoforms by way of transfection, and studies of the single-channel behavior of each specific isoform.

ACKNOWLEDGMENTS

Original work reported in this review was supported in part by grants from MURST 60% (PV) and MURST 60%, CNR grants 93.00301, CT04 and 94.2424, CT04 (LG).

REFERENCES

1. **Takeshima, H., Nishimura, S., Matsumoto, T., Ishida, H., Kangawa, N., Minamino, N., Matsuo, H., Ueda, M., Hanaoka, M., Hirose, T., and Numa, S.,** Primary structure and expression from complementary DNA of skeletal muscle ryanodine receptor, *Nature (London)*, 339, 439, 1989.
2. **Otsu, K., Willard, H. F., Khanna, V. K., Zorzato, F., Green, N. M., and MacLennan, D. H.,** Molecular cloning of cDNA encoding the Ca^{2+} release channel (ryanodine receptor) of rabbit cardiac muscle sarcoplasmic reticulum, *J. Biol. Chem.*, 265, 13472, 1990.
3. **Marks, A. R., Tempst, P., Hwang, K. S., Taubman, M. B., Inui, M., Chadwick, C., Fleischer, S., and Nadal-Ginard, B.,** Molecular cloning and characterization of the ryanodine receptor/junctional channel complex cDNA from skeletal muscle sarcoplasmic reticulum, *Proc. Natl. Acad. Sci. U.S.A.*, 86, 8683, 1989.

4. **Franzini-Armstrong, C. and Nunzi, G.,** Junctional feet and particles in the triads of a fast-twitch muscle fiber, *J. Muscle Res. Cell Motil.*, 4, 233, 1983.
5. **Inui, M., Saito, A., and Fleischer, S.,** Purification of the ryanodine receptor and identity with feet structures of junctional terminal cisternae of sarcoplasmic reticulum from fast skeletal muscle, *J. Biol. Chem*, 262, 1740, 1987.
6. **Zorzato, F., Fujii, J., Otsu, K., Phillips, M., Green, N. M., Lai, F. A., Meissner, G., and MacLennan, D. H.,** Molecular cloning of cDNA encoding human and rabbit forms of the Ca^{2+} release channel (ryanodine receptor) of skeletal muscle sarcoplasmic reticulum, *J. Biol. Chem.*, 265, 2244, 1990.
7. **Zhang, Y., Chen, H. S., Khanna, V. K., De Leon, S., Phillips, M. S., Schappert, K., Britt, B. A., Brownell, A. K. W., and MacLennan D. H.,** A mutation in the human ryanodine receptor gene associated with central core disease, *Nature Genet.*, 5, 46, 1993.
8. **Zorzato, F., Sacchetto, R., and Margreth, A.,** Identification of two ryanodine receptor transcripts in neonatal, slow-, and fast-twitch rabbit skeletal muscles, *Biochem. Biophys. Res. Commun.*, 203, 1725, 1994.
9. **Giannini, G., Clementi, E., Ceci, R., Marziali, G., and Sorrentino, V.,** Expression of a ryanodine receptor-Ca^{2+} channel that is regulated by TGF-ß, *Science*, 257, 91, 1992.
10. **Hakamata, Y., Nakai, J., Takeshima, H., and Imoto, K.,** Primary structure and distribution of a novel ryanodine receptor/calcium release channel from rabbit brain, *FEBS Lett.*, 312, 229, 1992.
11. **Giannini, G., Conti, A., Mammarella, S., Scrobogna, M., and Sorrentino, V.,** The ryanodine receptor/calcium channel genes are widely and differentially expressed in murine brain and peripheral tissues, *J. Cell Bio.*, 128, 893, 1995.
12. **Nakai, J., Imagawa, T., Hakamata, Y., Shigekawa, M., Takeshima, H., and Numa, S.,** Primary structure and functional expression from cDNA of the cardiac ryanodine receptor/calcium release channel, *FEBS Lett.*, 271, 169, 1990.
13. **Lesh, R. E., Marks, A. R., Somlyo, A. V., Fleischer, S., and Somlyo, A. P.,** Anti ryanodine receptor antibody binding sites in vascular and endocardial endothelium, *Circ. Res.*, 72, 481, 1993.
14. **Gorza, L., Vettore, S., Martini, A., Volpe, P., and Sorrentino, V.,** Temporal and regional differences in mRNA composition of intracellular Ca^{2+}-release channels of the heart, submitted results.
15. **Jorgensen, A. O., Shen, A. C. Y., Arnold, W., McPherson, P. S., and Campbell, K. P.,** The Ca^{2+} release channel/ryanodine receptor is localized in junctional and corbular sarcoplasmic reticulum in cardiac muscle, *J. Cell Biol.*, 120, 969, 1993.
16. **Arai, M., Otsu, K., MacLennan, D. H., and Periasamy, M.,** Regulation of sarcoplasmic reticulum gene expression during cardiac and skeletal muscle development, *Am. J. Physiol.*, 262, C614-C620, 1992.
17. **Wibo, M., Bravo, G., and Godfraind, T.,** Postnatal maturation of excitation-contraction coupling in rat ventricle in relation to the subcellular localization and surface density of 1,4-dihydropyridine and ryanodine receptors, *Circ. Res.*, 68, 662, 1991.
18. **Dutro, S. M., Airey, J. A., Beck, C. F., Sutko, J. L., and Trumble, W. R.,** Ryanodine receptor expression in embryonic avian cardiac muscle, *Dev. Biol.*, 55, 431, 1993.
19. **Takeshima, H., Nishimura, S., Nishi, M., Ikeda, M., and Sugimoto, T.,** A brain-specific transcript from the 3′-terminal region of the skeletal muscle ryanodine receptor gene, *FEBS Lett.*, 322, 105, 1993.
20. **Furuichi, T., Furutama, D., Hakamata, Y., Nakai, J., Takeshima, H., and Mikoshiba K.,** Multiple types of ryanodine receptor/Ca^{2+} release channels are differentially expressed in rabbit brain, *J. Neurosci.*, 14, 4794, 1994.
21. **Lai, F. A., Dent, M., Wickenden, C., Xu, L., Kumari, G., Misra, M., Lee, H. B., Sar, M., and Meissner, G.,** Expression of a cardiac Ca^{2+}-release channel isoform in mammalian brain, *Biochem. J.*, 288, 553, 1992.

22. **Nakanishi, S., Kuwajima, G., and Mikoshiba, K.,** Immunohistochemical localization of ryanodine receptors in mouse central nervous system, *Neurosci. Res.*, 15, 130, 1992.
23. **Kuwajima, G., Futatsugi, A., Niinobe, M., Nakanishi, S., and Mikoshiba, K.,** Two types of Ryanodine receptors in mouse brain: skeletal muscle type exclusively in Purkinje cells and cardiac muscle type in various neurons, *Neuron*, 9, 1133, 1992.
24. **Airey, J. A., Grinsell, M. M., Jones , L. R., Sutko, J. L., and Witcher, D.,** Three ryanodine receptor isoforms exist in avian striated muscles, *Biochemistry*, 32, 5739, 1993.
25. **Percival, A. L., Williams, A. J., Kenyon, J. L., Grinsell, M. M., Airey, J. A., and Sutko, J. L.,** Chicken skeletal muscle Ryanodine receptor isoforms: ion channel properties, *Biophys. J.*, 67, 1834, 1994.
26. **Oyamada, H., Murayama, T., Takagi, T., Iino, M., Iwabe, N., Mijata, T., Ogawa, Y., and Endo, M.,** Primary structure and distribution of Ryanodyne-binding protein isoforms of the bullfrog skeletal muscle, *J. Biol. Chem.*, 269, 17206, 1994.
27. **Airey, J. A., Beck, C. F., Murakami, K., Tanksley, S. J., Deerinck, T. J., Ellisman, M. H., and Sutko, J. L.,** Identification and localization of two triad junctional foot protein isoforms in mature avian fast-twitch skeletal muscle, *J. Biol. Chem.*, 265, 14187, 1990.
28. **Olivares, E. B., Tanksley, S. J., Airey, J. A., Beck, C. F., Ouyang, Y., Deerinck, J. T., Ellisman, M. H., and Sutko, J. L.,** Nonmammalian vertebrate skeletal muscles express two triad junctional foot protein isoforms, *Biophys. J.*, 59, 1153, 1991.
29. **O'Brien, J., Meissner, G., and Block, B.,** The fastest contracting muscles of nonmammalian vertebrates express only one isoform of the ryanodine receptor, *Biophys. J.*, 65, 2418, 1993.
30. **Walton, P. D., Airey, J. A., Sutko, J. A., Beck, C. F., Mignery, G. A., Sudhof, T. C., Deerinck, T. J., and Ellisman, M. H.,** Ryanodine and inositol trisphosphate receptors coexist in avian cerebellar Purkinje neurons, *J. Cell Biol.*, 113, 1146, 1991.
31. **Ellisman, M. H., Deerjnck, T. J., Ouyang, Y., Beck, C. F., Tanksley, S. J., Walton, P. D., Airej, J. A., and Sutko, J. L.,** Identification and localization of ryanodine binding proteins in the avian central nervous system, *Neuron*, 5, 135, 1990.
32. **Sudhof, T. C., Newton, C. L., Archer, B. T., III, Ushkaryov, Y. A., and Mignery, G. A.,** Structure of a novel $InsP_3$ receptor, *EMBO J.*, 10, 3199, 1991.
33. **Mignery, G. A., Newton, C. L., Archer, B. T., III, and Sudhof, T. C.,** Structure and expression of the rat inositol 1,4,5-trisphosphate receptor, *J. Biol. Chem.*, 265, 12679, 1990.
34. **Ross, C. A., Danoff, S. K., Schell, M. J., Snyder, S. H., and Ullrich, A.,** Three additional inositol 1,4,5-trisphosphate receptors: molecular cloning and differential localization in brain and peripheral tissues, *Proc. Natl. Acad. Sci. U.S.A.*, 89, 4265, 1992.
35. **Nakagawa, T., Okano, H., Furuichi, T., and Aruga, J.,** The subtypes of the mouse inositol 1,4,5-trisphosphate receptor are expressed in a tissue-specific and developmentally specific manner, *Proc. Natl. Acad. Sci. U.S.A.*, 88, 6244, 1991.
36. **Shell, M. J., Danoff, S. K., and Ross, C. A.,** Inositol (1,4,5)-trisphosphate receptor: characterization of neuron-specific alternative splicing in rat brain and peripheral tissues, *Mol. Brain Res.*, 17, 212, 1993.
37. **Furuichi, T., Simonchazottes, D., Fujino, I., Yamada, N., Hasegawa, M., and Miyawaki, A.,** Widespread expression of inositol 1,4,5-trisphosphate receptor type 1 gene (INSP3R1) in the mouse central nervous system, *Receptors Channels, 1,* 11, 1993.
38. **Nakagawa, T., Shiota, C., Okano, H., and Mikoshiba, K.,** Differential localization of alternative spliced transcripts encoding inositol 1,4,5-trisphosphate receptors in mouse cerebellum and hippocampus: *in situ* hybridization study, *J. Neurochem.*, 57, 1807, 1991.
39. **Sharp, A. H., McPherson, P. S., Dawson, T. M., Aoki, C., Campbell, K. P., and Snyder, S. H.,** Differential immunohistochemical localization of inositol 1,4,5-trisphosphate- and ryanodine-sensitive Ca^{2+} release channels in rat brain, *J. Neurosci.*, 13(7), 3051, 1993.

40. **Pozzan, T., Rizzuto, R., Volpe, P., and Meldolesi, J.,** Molecular and cellular physiology of intracellular calcium stores, *Physiol. Rev.,* 74, 595, 1994.
41. **Hain, J., Nath, S., Mayerleitner, M., Fleischer, S., and Schindler, H.,** Phosphorylation modulates the function of the calcium release channel of sarcoplasmic reticulum from skeletal muscle, *Biophys. J.*, 67, 1823, 1994.
42. **Meszaros, L. G., Bak, J., and Chu, A.,** Cyclic ADP-ribose as an endogenous regulator of the non-skeletal type ryanodine receptor Ca^{2+} channel, *Nature (London)*, 364, 76, 1993.
43. **Bezprozvanni, I., Watras, J., and Ehrlich, B. E.,** Bell-shaped calcium-response curves of Ins(1,4,5)P_3-gated and calcium-gated channels from endoplasmic reticulum of cerebellum, *Nature (London),* 351, 751, 1991.
44. **Iino, M. and Endo, M.,** Calcium-dependent immediate feed-back control of inositol 1,4,5-trisphosphate-induced Ca^{2+} release, *Nature (London)*, 360, 76, 1992.
45. **Berridge, M. J.,** Inositol trisphosphate and calcium signaling, *Nature (London)*, 361, 315, 1993.
46. **Llano, I., DiPolo, R., and Marty, A.,** Calcium induced calcium release in cerebellar Purkinje cells, *Neuron*, 12, 663, 1994.
47. **Verma, A., Hirsch, D. J., and Snyder, S. H.,** Calcium pools mobilized by calcium or inositol 1,4,5-trisphosphate are differentially localized in rat heart and brain, *Mol. Biol. Cell*, 3, 621, 1992.
48. **Bliss, T. V. P. and Collingridge, G. L.,** A synaptic model of memory: long-term potentiation in the hippocampus, *Nature (London)*, 361, 31, 1993.
49. **Gorza, L., Schiaffino, S., and Volpe, P.,** Inositol 1,4,5,-trisphosphate receptor in heart: evidence for its concentration in Purkinje myocytes of the conduction system, *J. Cell Biol.*, 121, 345, 1993.
50. **Fabiato, A.,** Calcium release in skinned cardiac cells: variation with species, tissues, and development, *Fed. Proc., 41,* 2238, 1982.
51. **Xu, L., Lai, F. A., Cohn, A., Etter, E., Guerrero, A., Fay, F. S., and Meissner, G.,** Evidence for a Ca^{2+}-gated ryanodine-sensitive Ca^{2+} release channel in visceral smooth muscle, *Proc. Natl. Acad. Sci. U.S.A.*, 91, 3294, 1994.
52. **Imagawa, T., Takasago, T., and Shigekawa, M.,** Cardiac ryanodine receptor is absent in type I slow skeletal muscle fibers: immunochemical and ryanodine binding studies, *J. Biochem. Tokyo*, 106, 342, 1989.
53. **Sorrentino, V.,** Personal communication.
54. **Gorza, L.,** Unpublished observation.

Chapter 8

RYANODINE RECEPTORS AND INTRACELLULAR CALCIUM SIGNALING

Michael J. Berridge, Timothy R. Cheek, Deborah L. Bennett, and Martin D. Bootman

CONTENTS

0-8493-8543-1/95/$0.00+$.50

1. INTRODUCTION

Cells have two major mechanisms available for regulating the release of internal calcium.[6,60] In one case, external signals acting on receptors at the cell periphery generate the second messenger inositol 1,4,5-trisphosphate ($InsP_3$) which diffuses into the cell and mobilizes stored calcium by engaging $InsP_3$ receptors ($InsP_3R$) on the endoplasmic reticulum.[6] The other mechanism employs a related family of intracellular channels, the ryanodine receptors (RyRs), so-called because they strongly bind the plant alkaloid ryanodine.[33,38,39,105,165] RyRs were first identified as the intracellular calcium channels responsible for releasing calcium from the sarcoplasmic reticulum of both skeletal and cardiac muscle. However, these receptors may be far more ubiquitous than imagined previously, as recent molecular and pharmacological evidence has revealed their presence in many different cell types. This chapter will attempt to bring together the emerging evidence indicating that RyRs of nonmuscle cells have a role to play in calcium signaling that may be equally important as the pathway involving $InsP_3Rs$.

2. CELLULAR DISTRIBUTION OF RYANODINE RECEPTORS

The distribution of RyRs in various tissues is still far from clear, but it is evident from Table 1 that these receptors are widespread. Like many other signaling components, it has become apparent that there is a family of RyRs. The two muscle types have been designated RyR1 (skeletal muscle) and RyR2 (cardiac muscle). A recently identified isoform, referred to as RyR3, appears to have a much more widespread distribution, including brain, smooth muscle, and some nonexcitable tissues.[58,165]

The distribution of RyRs has been studied extensively in the CNS.[82,120,133,157] Current evidence suggests that RyR2 is expressed throughout the brain,[82] particularly in the motor, visual, vestibular, olfactory bulb, and hippocampal regions.[133,157] RyR3 is also expressed in brain, but appears to have a more restricted distribution.[58] Expression of RyR1 in brain is predominantly detected only in cerebellar Purkinje cells.[82,133]

Another tissue with a complex pattern of RyR expression is smooth muscle.[112] Several lines of evidence suggest that smooth muscle cells express RyRs, including caffeine-induced calcium release[67,71,74,75,77,102,189] and ryanodine binding and modulation of calcium release.[67,75,77,96,187,189,193] Furthermore, RyRs from canine aorta[61] and toad visceral smooth muscle[189] have been partially purified and reconstituted into lipid bilayers, where they displayed a similar pharmacological profile to cardiac/skeletal RyRs.

In contrast to these observations, several studies have failed to obtain evidence supporting the expression of RyRs in smooth muscle (Table 1), for example, ryanodine binding was not detected using bovine aortic smooth

TABLE 1
Expression of RyR Isoforms in Various Tissues

		Detection Method						
		Pharmacological			Molecular			
						Genes		
Tissue	**Detected (+/-)**	**Ryanodine**	**Caffeine**	**cADPR**	**Ab**	**RyR1**	**RyR2**	**RyR3**
Skeletal muscle	+	13, 23, 50, 83, 84, 97, 108,116, 134, 141, 147	84, 134			99, 132, 172,195		
	–			110			119, 132	
Neuronal: Cerebellum: Purkinje	+	11, 93	11		82, 133, 157	82, 133	82	
	–							
Cerebellum: Granule cells	+	70	70		82		82	
	–							
Hippocampus	+		159		82, 157		82	58
	–					82		
Sensory neurons	+		176					
	–							

TABLE 1 (continued)
Expression of RyR Isoforms in Various Tissues

Tissue	Detected (+/-)	Detection Method						
		Pharmacological			Molecular			
						Genes		
		Ryanodine	Caffeine	cADPR	Ab	RyR1	RyR2	RyR3
Rat / bullfrog sympathetic neurons	+	43, 65, 100, 175	43, 65, 100, 175					
	–							
Cardiac muscle	+	2, 79, 109, 147, 178	26, 129, 178	110	79	99	115, 119, 132	
	–					132		
Chromaffin cells	+	19, 20, 98, 166	19, 20, 21, 92, 98, 166					
	–							
PC12 cells	+	3, 144, 192	3, 144, 192					
	–							
Exocrine pancreas	+	123, 177	123, 131, 155, 177, 180	177				
	–		76					
β Cells	+	170	146	170				
	–	146						

Smooth muscle	+	61, 67, 75, 77, 96, 187, 189, 193	61, 67, 74, 75, 77, 102, 189		189	51, 99	88, 119, 189	51, 58
	–	17	96			132, 189	132	
Hepatocyte	+	122, 152, 160, 161						
	–		153			99, 132, 161	132	
Endothelial cells	+	22, 194	12, 22, 52, 126		88		88	
	–					88		
Kidney	+							
	–					99, 132	119, 132	
Sea urchin eggs	+	14, 45, 46, 106	14, 45, 46, 86, 87, 106	45, 46, 87, 181	106			
	–							
Mouse eggs	+	169						
	–	114	114		114			
Xenopus oocytes	+							
	–	46, 138	46, 85	46	138	138		

Note: The table summarizes experimental evidence both for and against RyR expression in various cell types. The rows bearing '+' denote positive responses to either the pharmacological agents or identification with antibodies/ molecular techniques, while rows bearing '–' indicate the converse. Such negative responses must be treated with caution, since they may reflect the specificity of the probes which may detect one of the RyR isoforms, but not the others. Pharmacological identification of RyR relied on observations of specific [^{3}H]ryanodine binding, or the ability of caffeine, ryanodine or cyclic ADP ribose (cADPR) to mobilize calcium from intracellular stores, or to modulate agonist-stimulated calcium signals. The molecular detection methods include Northern blotting, Western blotting and polymerase chain reaction (PCR) analysis. Where isoform-specific probes were used for the molecular detection methods, the identified RyR isoform is indicated.

muscle microsomes[17] and caffeine did not induce calcium release from human myometrial cells,[96] A7r5 cells,[111] or porcine aortic smooth muscle.[73] These apparent discrepancies are probably explained by the fact that the level of RyR expression in smooth muscle varies considerably between tissues,[193] and is generally much lower than in striated muscle.[187] Alternatively, the lack of effect of caffeine on some smooth muscle cells may indicate the presence of RyR3, since this channel appears to be ryanodine sensitive, but caffeine insensitive.[49] In support of this, RyR3 expression has been identified in several smooth muscle tissues.[58]

The lack of sensitivity to caffeine may also be explained by the loss of RyR expression. There is growing evidence that the expression of RyRs may be somewhat labile in that these receptors soon disappear when certain cells are cultured.[51] For example, after 5 days in culture, aortic smooth muscle cells change from a 'contractile' to a 'synthetic' phenotype with a corresponding loss of caffeine sensitivity.[73, 102] A similar loss of responsiveness was detected in cultured cerebellar granule cells.[70]

These data suggest that RyR expression may be associated with the state of cell proliferation and differentiation. In the case of aortic smooth muscle cells, the caffeine-sensitive stores are lost during the logarithmic proliferation phase,[102] which can be reversed or augmented by removal or addition of growth factors, respectively. Expression of RyR1 in the BC_3H1 muscle cell line is prevented if cells are kept proliferating by growth factors, but RyR1 appears as the cells stop growing and begin to differentiate.[28,99] In the case of another cell line (Mv1Lu), the expression of RyR3 can be induced by transforming growth factor-β (TGF-β).[49]

The situation concerning the expression of RyRs in some nonexcitable tissues, and their participation in agonist-stimulated calcium signals, is far from clear. The data obtained using hepatocytes, for example, have produced conflicting results. Effects of ryanodine on agonist-induced calcium signals in intact hepatocytes[122,152] and specific ryanodine binding sites in hepatocyte vesicles[37,160,161] have been demonstrated. However, the pharmacology of the putative RyR in hepatocytes is very different from those expressed in muscle tissues,[107,153,161] and furthermore, the expression of RyR mRNA in hepatocytes has not been detected using molecular techinques.[99,132,161] These data suggest that if hepatocytes do express a ryanodine-binding protein, it is not the same as the RyR found in muscle and neuronal cells.

Experimental observations suggesting that the RyR is present in pancreatic acinar cells have also generated considerable interest recently. For example, elevation of the intracellular ($[Ca^{2+}]_i$) concentration by calcium infusion activates a calcium-induced calcium release (CICR) mechanism, indicative of RyR.[76,131,180] Additionally, calcium- and agonist-stimulated calcium signals are sensitive to the RyR activity modulators caffeine, ruthenium red, and ryanodine,[123,131,177] and can be stimulated by the putative RyR agonist cyclic adenosine diphosphate ribose (cyclic ADP ribose).[177] Furthermore, a caffeine-sensitive calcium channel was detected in the endoplasmic reticulum of pancreatic acinar cells, although this channel may not be responsible for the responses described above, since it was ryanodine and calcium insensitive.[155]

Another cell type with an interesting RyR expression is sea urchin eggs. Some eggs such as *Xenopus*, do not possess RyRs (Table 1), however, sea urchin eggs are sensitive to the RyR activators caffeine, ryanodine, and cyclic ADP ribose, suggesting that these cells express RyRs. Immunocytochemical analysis identified a protein surrounding the cortical granules and associated with the plasma membrane of mature eggs,[106] suggesting a role as an amplification mechanism for calcium signals occurring at fertilization. Interestingly, the immunoreactive protein identified in sea urchin eggs using the RyR1 antibodies was considerably smaller than the other known RyR isoforms. Additionally, photoaffinity labeling of cyclic ADP ribose-binding sites in sea urchin eggs identified two further proteins which were also much smaller than RyR1 and 3.[181]

The observations described above for hepatocytes, pancreas, and sea urchin eggs, where cells express RyR-like proteins with unusual pharmacology and molecular details, may point to the existence of novel RyR isoforms which remain to be identified.

3. SUBCELLULAR DISTRIBUTION OF RyR AND RELATIONSHIP TO $InsP_3$-SENSITIVE CALCIUM STORES

When considering the subcellular distribution of RyRs, it is also important to consider the functional relationship between RyR- and $InsP_3$-sensitive calcium stores. It seems that some cells exclusively express either RyRs (e.g., skeletal muscle) or $InsP_3$Rs (e.g., *Xenopus* and mammalian eggs). However, many other cells seem to have both receptor types, and there is much interest concerning the functional relationship of the two receptors.

Cardiac myocytes are an example where cells coexpress the RyR and $InsP_3$R, but the two receptors appear to have a clear functional distinction, since RyR2 are located on the sarcoplasmic reticulum at the triadic junctions, where they participate in excitation-contraction coupling, whereas the $InsP_3$Rs are located on stores clustered around the intercalated discs.[79] In snail neurons, the $InsP_3$-sensitive store is located near the nucleus, whereas the RyR-sensitive store lies close to the plasma membrane.[80] A similar localization of RyRs close to the plasma membrane occurs in brainstem and cerebellar neurons,[150] such an association would favor their role as an amplification mechanism of calcium signals entering from the outside.

In cerebellar Purkinje cells, $InsP_3$Rs and RyRs are found throughout the cell except in the spines, where only the former are located.[32,182] Elsewhere in the neuron the two receptors can either be located on separate stores or are colocalized.[182] A different arrangement seems to occur in the hippocampus, where RyRs were present in cell bodies, dendritic shafts, and spines, while $InsP_3Rs$ were present in cell bodies and shafts, but were not detected in the spines.[157]

A similarly complex arrangement, with partially overlapping ryanodine- and $InsP_3$-sensitive stores, has also been proposed for other neurons,[70] chromaffin cells,[166] PC12 cells,[3,192] and smooth muscle cells.[67,75,77] Interestingly, with smooth muscle cells, the extent of overlap between the two stores varies in different tissues,[67] and there may even be no overlap.[74] Presumably, these complex arrangements of intracellular calcium pools allow cells to uniquely control their responses to various extracellular stimuli.

Even when there is no physical overlap between the RyR and $InsP_3R$, it is clear that these two stores can be linked functionally in the sense that calcium released by one channel affects the activity of the other. This is most easily measured by examining the effect of $InsP_3$-mobilizing agonists in the presence of ryanodine, which can trap RyRs into an altered conducting state, indicating that they have been activated (Figure 1a and b). For example, stimulation of PC12 cells with ATP in the simultaneous presence of ryanodine inhibits the subsequent responses to ATP or caffeine, suggesting that calcium release from $InsP_3$-sensitive stores recruited RyR in these cells (Figure 1b).[3,144] Similar responses have been observed in smooth muscle cells,[75,77] bovine adrenal chromaffin cells,[166] and cerebellar granule neurons.[70] However, the recruitment of RyRs by $InsP_3$-mobilizing agonists is not a universal observation.[68]

This critical question, concerning how RyRs expressed in cells other than skeletal and cardiac muscle contribute to cellular calcium signaling, will be dealt with in greater detail later.

4. PROPERTIES OF RYANODINE RECEPTORS

Before considering how RyRs contribute to intracellular calcium signals, it is necessary to understand the regulation of these calcium channels. Much attention has been focused on the action of endogenous and exogenous factors responsible for regulating the opening and closing of the RyR. This section discusses some of the more important regulatory factors.

4.1. SENSITIVITY TO CYTOSOLIC CALCIUM

One of the major regulators of RyRs is calcium itself, which can both potentiate and inhibit calcium release. The phenomenon of CICR was first described for RyR2 of cardiac muscle,[34,41] and seems to be a general property of the ryanodine receptor family.[33,39,60] Some of the agents which influence RyR activity do so by altering the sensitivity of RyRs to cytosolic calcium.

RyRs display a bell-shaped activation curve; increasing cytosolic calcium in the low micromolar range stimulates channel opening, while further elevation of cytosolic calcium is inhibitory.[7,33,97] Such calcium activation curves are usually derived from steady-state measurements using permeabilized muscle

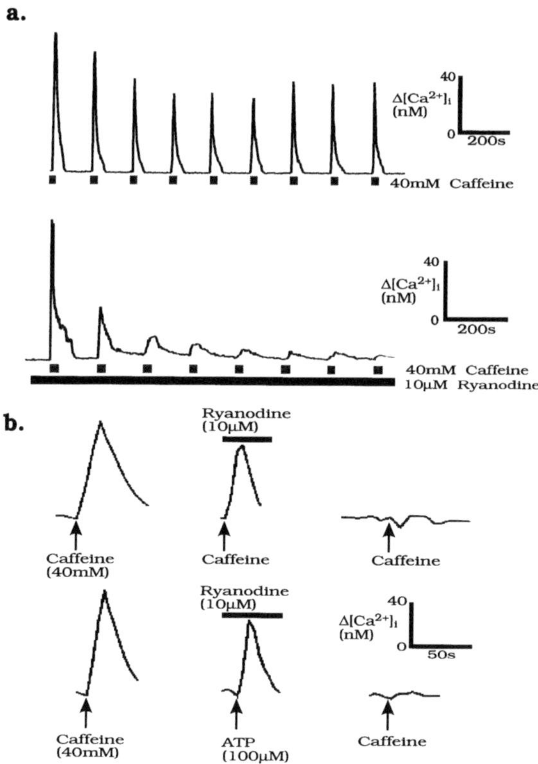

FIGURE 1. Activation and inhibition of calcium release from RyR by caffeine and ryanodine. **a.** and **b.,** Illustrate the use-dependent inhibition of caffeine- and agonist-induced Ca^{2+} release by ryanodine from a single bovine adrenal chromaffin cell (**a**), or a population of PC12 cells (**b**). The $[Ca^{2+}]_i$ was measured with Fura2, using methods described in Reference 20. Ryanodine had almost no effect when applied to either cell type on its own (not shown). **c.**, Illustrates graded calcium release and increase in caffeine-sensitivity following store loading. Caffeine was applied to a single Fura2-loaded bovine adrenal chromaffin cell incubated in calcium-free medium (**Ca^{2+}_0 free**). A transient $[Ca^{2+}]_i$ increase was observed during each stepwise increase in caffeine concentration (shown by the filled bars), suggesting that RyRs release calcium from the intracellular pool in a graded manner. Once the caffeine-sensitive pool was emptied, it could be rapidly refilled by returning the cells to a calcium-containing medium (3 m*M* Ca^{2+}_o), followed by depolarization with nicotine (**N**; 10 μ*M*), which evoked a large $[Ca^{2+}]_i$ elevation. In all cells tested ($n = 8$), depolarization induced a greater loading of the caffeine-sensitive calcium pool than in nondepolarized cells. Concomitant with the elevated calcium pool loading, the cells were more sensitive to caffeine when tested again with the same caffeine concentrations, as shown. The increased sensitivity to caffeine suggests that lumenal calcium may control the activity of RyR in these cells. Similar data have been presented in Reference 20.

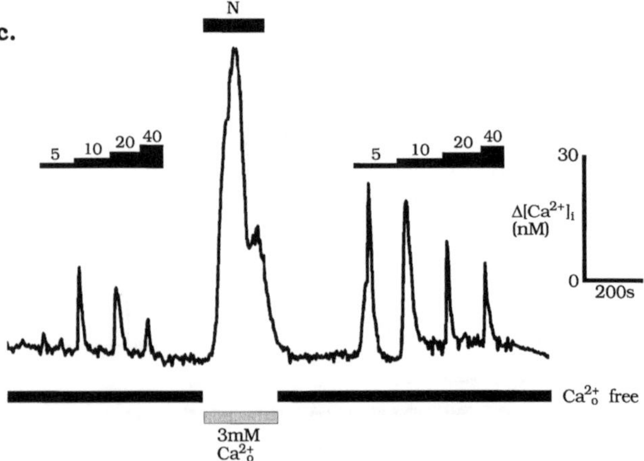

FIGURE 1 (continued).

fibers, SR vesicles, or purified RyRs incorporated into lipid bilayers. This experimental approach may have little physiological significance because the level of cytosolic calcium is rarely sustained at different levels, but is usually presented in the form of short-lived rapid elevations. Indeed, cardiac muscle RyRs were more substantially activated by a rapid calcium increase than by a slower increase of a similar magnitude.[35] A comparable observation was made using single cardiac RyR channels incorporated into lipid bilayers, where the probability of channel opening caused by photolysis of caged calcium was initially higher than during the subsequent steady state.[56]

Such responses may indicate a difference in the rate of association between calcium and the binding sites that mediate either the activation or inhibition of RyRs. This mechanism could help to reject spurious triggering events and ensure that RyR activation always precedes inactivation. However, it should be mentioned that the case for time-dependent RyR inactivation is not clear-cut. Although some studies have detected a decrease in RyR channel activity following elevation of cytosolic calcium concentration,[35,56,117,156] other studies do not support time-dependent RyR inactivation.[33]

4.2. RYANODINE BINDING

The name of the RyR is derived from its high affinity for the plant alkaloid, ryanodine, which has effects on channel opening and has been used as a probe during RyR purification.[2,61,189] Apart from being a specific ligand for RyRs, ryanodine has several properties which make it extremely useful in studies of calcium signaling. Firstly, the effect of ryanodine on RyR channel activity is concentration dependent; at nanomolar concentrations ryanodine reversibly increases the probability of channel opening, at around 0.1 to 10 μM, ryanodine promotes a long-lived subconductance state, while higher concentrations irreversibly close the channel.[2,83,106,116,147] In the open-locked subconductance state, RyRs remain sensitive to the modulatory effects of calcium and adenine nucleotides,[134] but not the channel inhibitor ruthenium red.[83,116,147]

Measurements of [^{3}H]ryanodine binding in several tissues have identified single high-affinity binding sites with dissociation constants (K_d) generally <10 n*M*.[2,104,193] However, more sophisticated analyses have indicated multiple (up to four) binding sites for ryanodine with K_d values ranging from 1 n*M* to 5 μ*M*.[13,136,141] The affinities of these binding sites correlated with observed changes in RyR channel activity at various ryanodine concentrations,[13] suggesting that the various effects of ryanodine are mediated by multiple allosteric binding sites.

The binding of ryanodine to its receptor is enhanced by addition of calcium, caffeine, adenine nucleotides, and increasing ionic strength.[25,108,109,136,140] Since these are experimental conditions that promote RyR channel activity, it is generally assumed that ryanodine binds more avidly to open RyRs.[108,134,140] This 'use-dependent' binding of ryanodine is a particularly advantageous property, since it provides a memory of previously activated RyRs. For this reason, ryanodine has been used to demonstrate the potential participation of RyR in calcium signaling during stimulation of a variety of cell types, including smooth muscle cells,[67,71] neurons,[11,43,70,158,175] cardiac cells,[127] chromaffin cells,[19,20,166] and PC12 cells.[3]

4.3. GRADED CALCIUM RELEASE

An interesting feature of calcium release from RyRs in intact cells is that it can be graded in direct proportion to the stimulus intensity. Such graded responses have been evident for some time in cardiac myocytes, where the amount of calcium released from the sarcoplasmic reticulum is graded depending on the extent of depolarization, which determines the magnitude of trigger calcium.[15,26,57,168] Graded responses to varying levels of depolarization have also been demonstrated using sympathetic ganglion neurons[65] and smooth muscle cells.[53] Additionally, graded responses have been observed using various neuronal cells stimulated with caffeine, where the extent of calcium release is proportional to the caffeine concentration (Figure 1c).[19,20,43,70] The opposite result, namely an all-or-none $[Ca^{2+}]_i$ response, where the response magnitude was independent of caffeine concentration, was obtained using Chinese hamster ovary cells transfected with rabbit skeletal RyR1.[139] The reasons underlying the difference between this all-or-none and other graded responses is not clear. However, as the observations of graded responses were made using cells that probably express RyR2, these data may indicate an interesting difference in the activation of RyR1 and RyR2.

Graded responses such as those described above are surprising, because RyRs display the autocatalytic process of CICR. The predicted response of a single cell, following the activation of a limited number of RyRs, would be the progressive activation and recruitment of other RyRs, leading to an all-or-none response. Stern[168] suggested that graded responses could not be obtained in cells where RyR with homogeneous sensitivities to trigger calcium, released their calcium into a common cytosolic pool.

The observation of graded responses using cardiac myocytes has been accounted for by a mechanism of 'local-control' where individual voltage-operated calcium channels (VOCs) are coupled to clusters of RyRs by means of a 'calcium synapse'.[168,184,186] It is suggested that as the plasmalemma becomes more depolarized, the opening of additional VOCs in the membrane recruits progressively more RyR clusters. These RyR clusters therefore represent functionally independent calcium release units which serve to amplify the trigger calcium signal, but they neither share this trigger calcium or coactivate each other. Paradoxically, these functionally independent units can also become coupled under certain conditions, for example during propagation of a calcium wave caused by SR calcium overload, probably because increasing lumenal calcium sensitizes RyRs (see Section 4.4.2) and will also facilitate the diffusion of calcium from one release unit to the next.

The mechanism underlying the graded responses following caffeine stimulation of various neurons[19,20,43,70] may have some similarity with the RyR cluster model in cardiac cells. The graded caffeine-stimulated signals in adrenal chromaffin cells (Figure 1c) seem to be most easily explained by a scheme in which increasing caffeine concentrations progressively recruit more RyRs (or RyR clusters). These RyR clusters may deplete their local calcium pool in an all-or-none manner, however, they do not activate neighboring channels, and can therefore be considered as functionally discrete units.[19,20] The progressive recruitment of these units by increasing caffeine concentrations suggests that they must have differing sensitivities either to caffeine, or calcium. The mechanism underlying these differential sensitivities is unknown, however, there are several factors that could control the sensitivity of RyRs, such as phosphorylation or a variation of lumenal calcium, as described below.

Additional evidence for functionally independent RyRs includes observations of spontaneous membrane potential fluctuations resulting from punctate calcium release beneath the plasma membrane of various cells,[4,9,100,103,154] and confocal imaging of single cardiac myocytes, where spontaneous highly localized calcium signals described as 'calcium sparks' have been observed (Figure 2a).[24,89] Graded responses may therefore reflect the integrated activity of those RyRs recruited by each particular level of stimulation.

These observations of graded responses by RyRs indicate an important similarity between RyRs and $InsP_3Rs$, since graded (or 'quantal') responses have also been found for hormone-induced calcium release via $InsP_3Rs$.[10,174] Furthermore, confocal imaging experiments have suggested that $InsP_3Rs$ may also operate in a functionally independent manner.[137,190] These data suggest that the fundamental calcium release units may be independent clusters of RyRs or $InsP_3Rs$, and provide valuable clues concerning the events leading up to the initiation of a global calcium signal, as described in a later section.

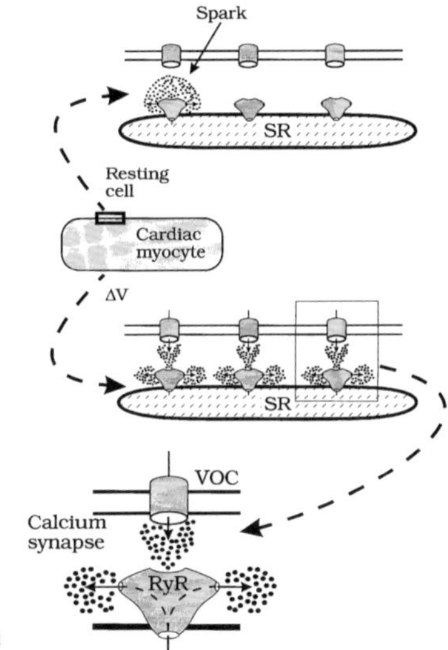

FIGURE 2. Mechanisms of RyR activation leading to calcium release and calcium entry. **(a)** This figure depicts two different kinds of cardiac RyR activity;[184] uncontrolled random RyR openings leading to calcium sparks (top) and controlled RyR activity resulting from electrical depolarization (Δ**V**) (middle). In both cases, the $[Ca^{2+}]_i$ rise resulting from RyR activation is localized and does not recruit neighboring RyRs. The bottom diagram is an enlargement of the proposed functional arrangement between voltage-operated calcium channels (**VOC**) in the sarcolemma and RyRs in the sarcoplasmic reticulum. In this scheme, membrane depolarization triggers an influx of calcium which crosses the gap between the two membranes and activates the RyRs, thus forming a 'calcium synapse'.[168] Cardiac $[Ca^{2+}]_i$ responses can be graded due to the progressive recruitment of these synapses as the membrane becomes increasingly depolarized.

4.4. MODULATION OF CHANNEL SENSITIVITY

The sensitivity of RyRs can be modulated by a variety of agents both endogenous and exogenous. As will become apparent, some of these agents seem to act by adjusting the sensitivity of the RyR to calcium.

4.4.1. Caffeine

Caffeine has been used extensively as a pharmacological agent to stimulate the opening of RyRs. The activity of channels treated with caffeine displays a bell-shaped dependence on cytosolic calcium, similar to that shown by untreated channels. However, caffeine dramatically increases the sensitivity of RyRs to calcium and also prolongs the duration of open channel events, without changing the unit conductance.[33,117,118,140,148] The ability of caffeine to release calcium in intact cells is therefore most probably due to caffeine sensitizing RyRs to the ambient $[Ca^{2+}]_i$.

The three RyR isoforms appear to have different sensitivities to caffeine, with cardiac RyR2 being the most sensitive isoform, followed by RyR1,

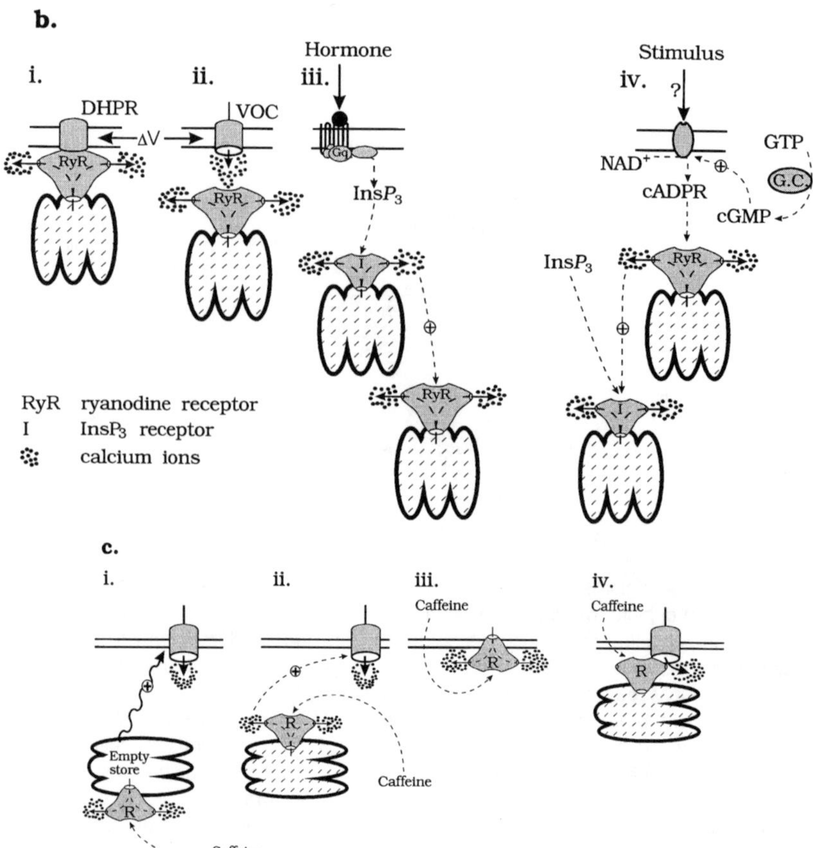

FIGURE 2 (continued). (b) The schemes presented in **i** to **iv** represent the various mechanisms that have been proposed to account for RyR (**RyR**) activation in different cell types. **i**, Conformational coupling, induced by membrane depolarization (ΔV), which is predominant in skeletal muscle. **ii**, Calcium-induced calcium release (CICR) triggered by calcium influx. **iii**, CICR triggered by internal calcium mobilization through $InsP_3Rs$ (**I**). **iv**, Calcium release triggered by cyclic ADP ribose (**cADPR**), and its proposed amplification by recruiting $InsP_3Rs$. **G.C.** and **cGMP** denote guanylate cyclase and cyclic guanosine monophosphate, respectively. **(c)** The diagrams represented in **i** to **iv** describe the possible mechanisms by which activation of RyRs (**R**) could lead to entry of calcium into cells. **i**, 'Capacitative' calcium entry caused by depletion of the intracellular calcium stores. **ii**, Calcium-activated calcium entry, where the increase in $[Ca^{2+}]_i$ opens a calcium-sensitive channel at the plasma membrane. **iii**, Direct gating of calcium into cells by RyR inserted into the plasma membrane. **iv**, Conformation coupling, this is analogous to the conformational coupling situation in skeletal muscle (see Figure **2b, i**), except that in this case, the information flows from the activated RyR to the channel in the membrane.

whereas RyR3 seems to be caffeine insensitive.[165] The insensitivity of RyR3 to caffeine suggests that caffeine is a poor probe for the presence of RyRs, and probably accounts for the failure to recognize the expression of RyRs in tissues where they have been found using molecular techniques.[49,58]

4.4.2. Lumenal Calcium

There is growing evidence that RyR1 and RyR2 may be sensitive to the calcium concentration within the sarcoplasmic reticulum.[50,125,164] The role of lumenal calcium is complex in that it modulates the response to cytosolic activators such as calcium and ryanodine, suggesting that lumenal calcium allosterically regulates the sensitivity of binding sites on the cytosolic surface of RyRs.

This phenomenon is not restricted to RyRs of muscle cells, but has also been suggested for the calcium stores in bovine chromaffin cells[18,20] (Figure 1c), and sea urchin eggs.[46] Additionally, the caffeine-sensitive calcium store in sympathetic ganglion neurons[43,91] and portal vein smooth muscle cells[53] can strongly influence the effect of membrane depolarization, depending on their state of filling; when the stores are full they can amplify calcium signals due to CICR, but they will blunt such responses if they are empty, by acting as calcium sinks. Modulation of RyR activity by lumenal calcium represents another similarity between RyRs and $InsP_3Rs$, since it has been demonstrated that the latter may also be regulated by lumenal calcium (reviewed in References 10 and 113).

How might the RyR sense the level of calcium within the lumen? Since the segments of the RyR that face the lumen lack obvious calcium-binding motifs, it has been proposed that the receptor is being altered by some other calcium-sensing molecule such as calsequestrin[69] or possibly annexin VI.[30] When added to the *trans* (lumenal) side of a bilayer containing RyRs, calsequestrin induced a large increase in channel activity.[78]

4.4.3. ATP

Like their $InsP_3R$ counterparts, RyRs are sensitive to ATP in the millimolar range that is normally found within cells. In fact, a variety of adenine nucleotides can modulate RyR channels activity, with the adenine moiety seemingly the most important part, as other nucleotides such as GTP, ITP, etc., are virtually ineffective.[33] The primary sequence of the RyR1 revealed several putative nucleotide-binding sequences.[172] ATP modulates the opening of RyRs, not by altering their sensitivity to cytosolic calcium, but rather by increasing the rate of calcium release.[33,140,148]

The functional significance of this allosteric regulation through ATP is unclear, but it may represent a protective device to prevent the stores from dumping their calcium should the cell run out of ATP.[8] During heart failure or stroke, the ischemia results in a rapid decrease in ATP concentration, which will attenuate the activity of RyRs and $InsP_3Rs$, thus reducing the danger of calcium toxicity.

4.4.4. Modulation Through Phosphorylation

RyRs contain a number of putative phosphorylation sites which may serve to modulate channel activity. In the case of cardiac muscle, CAM kinase is tightly associated with RyR2 [63] where it acts to phosphorylate serine 2809 resulting in channel activation.[188] On the other hand, phosphorylation of RyR1 in skeletal muscle by CAM kinase-II results in an increased sensitivity to calcium.[183] The brain RyR can be phosphorylated by cyclic AMP-dependent protein kinase, resulting in an increase in sensitivity to caffeine.[191]

5. RyR FUNCTION IN CALCIUM SIGNALING

Having dealt with the main properties of RyRs, we are now in a position to consider how they might contribute towards calcium signaling in different cell types. While their role in excitation-contraction coupling in muscle cells is well established, we are just beginning to appreciate that they may play an equally important signaling role in many nonmuscle cells.

5.1. RELEASE OF INTERNAL CALCIUM

One of the primary functions of RyRs is to release calcium from internal reservoirs, so the question we have to consider is how the external signal detected at the cell surface can activate RyRs embedded in the internal stores. In the case of $InsP_3Rs$, exactly the same problem is solved by using the diffusible second messenger $InsP_3$.[6] While a similar strategy may also apply to some RyRs, the situation is clearly more complex because the opening of these receptors can be controlled by different mechanisms.

5.1.1. Conformational Coupling

In skeletal muscle, release of calcium from the sarcoplasmic reticulum (SR) is known to be triggered by membrane depolarization as the action potential invades the T-tubules.[31] Since it is impossible for current to cross the 20 n*M* gap separating the plasmalemma and sarcolemma, because cytosolic ions will rapidly dissipate the charge, there is a need for some "chemical messenger" to communicate between the two membranes. A widely supported hypothesis is that RyR1 itself might be the messenger transmitting information from the plasma membrane to the SR.[145] Structural studies have revealed protein strands, the T-tubule feet, linking the two membranes.[42] These foot structures are now thought to be the bulbous heads of the RyR1 which are sufficiently large to span the 20 n*M* gap and to make contact with the dihydropyridine receptor (DHPR) (Figure 2b, i).[83,172,173,195] The latter is the voltage sensor that detects membrane depolarization and undergoes a conformational change which is then transmitted to the RyR1, resulting in an opening of the calcium channel. An association between these two proteins has been demonstrated in immunoprecipitation experiments.[101] The bulbous head of the RyR therefore functions as a molecular transduction unit to transfer information from the DHPR to open its calcium channel in the SR membrane (Figure 2b, i).

The cytoplasmic loop between the second and third transmembrane repeats of the DHPR $\alpha 1$ subunit seems to be responsible for the functional interaction with the RyR[95,173] since addition of just this segment of the DHPR $\alpha 1$ subunit can activate skeletal RyRs. Interestingly, the cytoplasmic loop between the second and third repeats of either the cardiac or skeletal DHPR $\alpha 1$ subunit can activate the purified skeletal mucle RyR. However, the same proteins cannot activate cardiac RyRs.[95] It seems, therefore, that RyR1 can act via a conformational coupling or CICR mechanism, but RyR2 can only act as CICR channels. This suggestion is consistent with the ability of either skeletal or cardiac DHPR $\alpha 1$ subunits to restore excitation-contraction coupling in dysgenic skeletal muscle cells.[1]

So far, this conformational coupling mechanism has only been described in skeletal muscle. However, it is possible that such a mechanism might occur in other cell types such as neurons, some of which have RyR1. Ultrastructural studies have revealed that the endoplasmic reticulum in neurons[59] and smooth muscle[29] is sometimes closely aligned to the plasma membrane, as it is in striated muscle, even to the extent of having regular protein connections between the two membranes. The similarity between neurons and muscle raises the possibility that neurons might release calcium from internal stores using a mechanism similar to that in skeletal muscle cells.[59] However, there is no evidence, as yet, that neurons can mobilize internal calcium by conformational coupling.

5.1.2. Calcium-Induced Calcium Release in Cardiac Cells

As described earlier, a major regulator of RyRs is calcium itself. The process of CICR is thus a key element in the activation of RyRs in many cell types. The classical example is cardiac muscle where VOCs in the sarcolemma gate a small pulse of calcium that triggers a regenerative calcium release from the RyR2 located in the SR (Figures 2a and 2b, ii).[168] It is difficult to detect the small pulse of trigger calcium, because it is rapidly swamped by the much larger bolus of calcium flooding out of the SR. However, the two calcium sources can be distinguished if the cardiac cell is injected with citrate, which disrupts the coupling between the small influx of trigger calcium and the release of internal calcium stores.[90] Another interesting observation concerning these experiments is that the effectiveness of the trigger calcium was critically dependent on the extent of store loading; if the stores were empty, the pulse of trigger calcium rapidly dissipated without causing release, but as the stores refilled a threshold was reached, when the same small pulse began to activate the RyRs. This may be consistent with the effect of lumenal calcium described above. Nevertheless, the process of CICR in normal cardiac myocytes does not become regenerative, in the sense that the opening of a few RyRs does not recruit neighboring channels to produce an all-or-none response. As described earlier, the release process in cardiac cells is graded depending on the degree of membrane depolarization.

Some insights into how calcium release might be graded have emerged from recent studies on ‘calcium sparks’, which may represent the unitary events associated with the opening and closing of a single RyR2 channel (Figure 2a).[24] Under normal conditions, isolated myocytes displayed brief highly localized $[Ca^{2+}]_i$ increases, which occurred randomly and at low frequency throughout the cell. These sparks were stochastic events resulting from the opening of either a single RyR, or clusters of RyRs. If the myocytes were placed in a high calcium medium, which increased the calcium load of the stores, spark frequency was greatly increased. Furthermore, the amplitude and area of some sparks enlarged to become macrosparks as they began to recruit neighboring receptors, thus forming an initiation site capable of spawning a regenerative calcium wave.[24]

Such calcium waves, observed in overloaded muscle, are probably not present during the normal process of E-C coupling.[185] However, these sparks,[24] and other observations of local $[Ca^{2+}]_i$ elevations which fail to cause regeneration,[89,130] give important clues to the normal process responsible for activating calcium release in the heart. Such localized responses indicate that the RyRs are capable of releasing calcium but do not initiate a regenerative wave, because they fail to activate their neighbors. The failure to induce regenerative activity is probably because induction of CICR requires a high calcium concentration in the vicinity of each RyR. For example, flash photolysis studies using caged calcium indicated that the threshold for CICR in crayfish muscle was above 0.9 μ*M*,[57] and in skinned barnacle muscle concentrations up to 10 μ*M* failed to activate CICR.[84] These data suggest that RyRs need to be bathed in a high concentration of calcium before they open.

The role of the action potential, therefore, is to coordinate the release activity of all the RyRs, by simultaneously introducing high-concentration pulses of trigger calcium through the VOCs distributed over the cell surface (Figure 2a). Graded release is achieved because the number of VOCs recruited by a depolarizing pulse is greater at more depolarizing potentials.[16] The opening of each VOC thus creates a microdomain of elevated $[Ca^{2+}]_i$, which will only activate those RyRs lying in the immediate vicinity of the calcium channel, because the $[Ca^{2+}]_i$ gradient steeply declines with distance from the VOC.[167]

It seems, therefore, that the process of CICR in normal cardiac cells depends upon a tight spatial coupling between VOCs and RyRs (i.e., the calcium synapse described in Figure 2a), and this accounts for how the release process can be graded depending upon the degree of membrane depolarization. Regenerative calcium waves may only occur under extreme conditions, i.e., when the SR is heavily loaded with calcium.

5.1.3. Calcium-Induced Calcium Release in Other Cell Types

The CICR mechanism (Figure 2b, ii) is not restricted to cardiac muscle but has also been described in a number of other cell types. There appear to be two separate mechanisms for generating the pulse of trigger calcium, it may be due to the entry of calcium (as in cardiac cells; Figure 2b, ii) or it may be released from internal stores (Figure 2b, iii). Examples of cells that employ the cardiac-type mechanism are smooth muscle cells,[48,53,71] GH_4C_1 pituitary cells,[179] and various neurons.[43,64,65,81,91,93,151] The evidence that membrane depolarization induces the entry of calcium, which then activates CICR in these tissues, is as follows:

1. Removal of extracellular calcium prevents internal release, arguing for CICR and against the conformational coupling mechanism used in skeletal muscle. Additionally, the concentration of external calcium surrounding spinal neurons can be titrated such that the influx component cannot be detected by voltage-clamp measurements, but without altering the depolarization-induced elevation of intracellular calcium. These data support the notion that the bulk of the calcium signal can be derived from the release of internal stores.[64]
2. The level of cytosolic calcium continued to increase after the depolarizing pulse was terminated in smooth muscle[53] and in sympathetic neurons,[62,65,91] suggesting that the CICR process continued to regenerate itself even when the entry of external calcium was curtailed. In the case of sympathetic neurons, the calcium signal begins immediately below the plasma membrane, reflecting the entry of external calcium, but then spreads inwards as a regenerative wave moving at 48 μm/s.[65]
3. Another indication that CICR contributes to depolarization-induced calcium signals has come from studying the relationship between the amplitude of the cytosolic calcium signal and the duration of the depolarizing stimulus.[65,93] If the calcium signal was solely due to calcium influx, a linear relationship would be expected. However, the relationship was found to be supralinear in both sympathetic ganglion neurons[65] and cerebellar Purkinje cells;[93] as the length of the depolarizing pulse increases, a greater proportion of the signal was derived from the internal stores. This supralinear response became linear when the sympathetic neurons were treated with ruthenium red, an inhibitor of RyR calcium release.
4. Manipulating the sensitivity of internal calcium stores can markedly influence the extent of depolarization-induced calcium signals. For example, depleting the internal stores with caffeine completely suppressed K^+-dependent calcium signaling in bullfrog sympathetic neurons.[43,91] Conversely, low ryanodine concentrations markedly potentiated calcium signaling in cerebellar Purkinje cells,[93] and low caffeine concentrations enhanced the depolarization-induced release of calcium in smooth muscle cells.[53] Additionally, perfusing the putative RyR agonist, cyclic ADP ribose, into sympathetic neurons also potentiated depolarization-induced $[Ca^{2+}]_i$ increases.[66]
5. The use-dependence of ryanodine binding to RyR has provided further evidence that these receptors are opened by depolarization-induced calcium influx. For example, a brief depolarization of bullfrog sympathetic neurons with a high potassium concentration, in the simultaneous presence of ryanodine, inhibited the subsequent response to caffeine.[43] The insensitivity of these open-locked RyRs to caffeine indicates that the receptors were opened during the depolarizing pulse.

The evidence described above provides a strong case for proposing that the RyRs of smooth muscle cells and neurons can contribute to calcium signals during membrane depolarization. The hypothesis is that a small entry of calcium through VOCs is amplified by CICR from stores containing RyRs, as was described for cardiac cells (Figures 2a and 2b, ii). In the case of smooth muscle, the release of internal calcium probably contributes to the depolarization-induced contraction. In neurons, the release of calcium from RyR-sensitive calcium stores may contribute to long-lasting afterhyperpolarizing (AHP) responses.[11,151] The ability of caffeine to modulate the firing pattern of cerebellar Purkinje cells may be mediated through an alteration of this AHP response.

However, it should be mentioned that this internal amplification process is not always present; for example, ryanodine had no effect on tetanus-induced calcium transients in freshly isolated bullfrog sympathetic neurons.[128] Additionally, responses observed using rat sympathetic neurons appear to differ from those described above for bullfrog sympathetic neurons, in that the caffeine-sensitive store is rather small and produces little amplification of depolarization-induced calcium entry.[175] A similar lack of amplification of depolarization-induced calcium signals by RyRs was reported for bovine adrenal chromaffin cells.[124]

5.1.4. Role of Internal Calcium as a Trigger for CICR

Other cell types employ an alternative mechanism for providing trigger calcium. In cells which have both $InsP_3$- and RyR-sensitive stores, the opening of $InsP_3Rs$ may provide the trigger calcium used to activate RyRs (Figure 2b, iii), as has been proposed for bovine chromaffin cells,[166] some smooth muscle cells,[149] murine macrophages,[143] cerebellar granule cells,[70] PC12 cells,[3,144] exocrine pancreatic cells,[123,131,180] hippocampal CA1 neurones,[158] and perhaps also in endothelial cells.[194] In endothelial cells, the entry of calcium through stretch-activated channels might be prolonged by secondary release from internal stores regulated by RyRs, using the mechanism described earlier (Figure 2b, ii).[121]

What is difficult to determine for any of these examples is just how the RyRs contribute to the agonist-induced calcium signals. The mechanism considered most often is that RyRs act to amplify the $InsP_3$-induced calcium response.[5] However, since $InsP_3Rs$ are sensitive to $[Ca^{2+}]_i$,[7] it is possible that the opposite might occur, i.e., RyR-released calcium enhances the activity of $InsP_3Rs$.[177]

5.1.5. Calcium Release by Cyclic ADP Ribose

In all the previous examples (Figure 2b, i to iii), RyRs were activated by calcium derived from either external or internal sources. Another possibility is that release might be modulated or triggered directly by means of the putative second messenger cyclic ADP ribose.[44] By analogy with $InsP_3$, the idea would be that an external stimulus would promote an increase in the level of cyclic ADP ribose which would then act on the RyR to initiate the release of calcium (Figure 2b, iv). Like the $InsP_3Rs$, which are sensitive to $InsP_3$ and calcium,

RyRs might be under dual agonist control because cyclic ADP ribose was found to sensitize the process of CICR in sea urchin homogenates.[87]

Sitsapesan et al.[163] have suggested that cyclic ADP ribose may release calcium by interacting with the ATP-binding site of RyRs. In this case, cyclic ADP ribose may not open the channel itself, but mimics the action of adenine nucleotides[33] (see Section 4.4) and simply potentiates channel opening at permissive calcium concentrations. In contrast, however, it has been demonstrated that cyclic ADP ribose stimulated RyR channel activity at low calcium concentrations, using either cardiac or sea urchin RyRs, suggesting that it may sensitize RyRs to cytosolic calcium in a similar manner to caffeine.[87,110]

The data presented by Sitsapesan et al.[163] also suggest that in the presence of physiological levels of ATP, cyclic ADP ribose may act as an antagonist of RyR activity, since it can only partially mimic the effect of ATP. These data imply that calcium-releasing effects of cyclic ADP ribose might be mediated by a channel other than the cardiac-type RyR, or that intermediary proteins are required, as suggested by Lee and colleagues.[87,181] More information is clearly required regarding the action of cyclic ADP ribose on RyRs to confirm its role as a physiological RyR activator.

Additionally, to fully establish a second messenger role for cyclic ADP ribose more evidence is needed that external stimuli act to increase the level of this messenger. There is some information that cyclic ADP ribose increases in β-cells in response to glucose,[170] and in sea urchin eggs, cyclic ADP ribose formation may be activated by cyclic GMP.[47] In neither case, however, is there any evidence that this messenger is generated in response to activation of classical signaling pathways such as the G protein-linked or tyrosine kinase-linked receptors. With regard to the latter mechanism, it has been proposed that cyclic ADP ribose might mediate the *Lyn*-dependent increase in calcium following activation of the B-cell antigen receptor.[171] There clearly is a need for more information on the factors responsible for regulating the intracellular level of cyclic ADP ribose and how this might vary during cell stimulation.

Pancreatic acinar cells appear to have both $InsP_3$- and ryanodine-sensitive stores that seem to interact with each other. When given alone, a low concentration (1 m*M*) of caffeine has little effect, but it can induce spiking when given together with a subthreshold level of acetylcholine[131] or during internal infusion with 100 μm calcium.[180] Higher doses of caffeine, or treatment with ryanodine, slowed down the velocity of the apical-basal waves induced by acetylcholine.[123] All this evidence seemed to be consistent with the notion that $InsP_3$-sensitive stores in the apical region initiate a calcium signal that is then spread through the cell by recruiting both RyRs and $InsP_3Rs$.[76,123,180] A role for RyRs has been further supported by recent studies on pancreatic acinar cells showing that cyclic ADP ribose was able to duplicate the normal agonist-induced calcium spiking patterns.[177] In addition, the stimulatory effects of acetylcholine and cholecystokinin were reduced or blocked by ryanodine and ruthenium red. These experiments raise yet another possible functional interaction between the different calcium stores, i.e., receptor activation of cyclic ADP ribose may release calcium from RyRs which is then further amplified by CICR using the $InsP_3Rs$ (Figure 2b, iv).[177]

5.1.6. Calcium Sparks and Initiation of Calcium Signals

The calcium sparks described earlier in cardiac cells (Figure 2a) may turn out to be a general property of both RyRs and $InsP_3Rs$ with important implications for the initiation of calcium signals. Indirect evidence for calcium sparking in both smooth muscle cells and neurons has come from the recording of spontaneous transient outward currents (STOC) in vascular and intestinal smooth muscle cells[4,9] and spontaneous miniature outward currents (SMOC) in dorsal root ganglion cells,[103] bullfrog sympathetic neurons,[100,154] and chick ciliary ganglion.[40] Both STOCs and SMOCs result from the localized release of calcium from an intracellular store which then activates approximately 50 to 1000 calcium-sensitive potassium channels to give these transient outward currents. The kinetics of these smooth muscle and neuronal sparks resemble those in cardiac cells — they have a very rapid rising phase as calcium rushes out of the channel, followed by a slower recovery phase as the microdomain of calcium dissipates by diffusion once the channel closes. The frequency of these sparks is greatly accelerated by caffeine in both smooth muscle cells[4,9] and neurons.[100,154] It is particularly interesting that an increase in spark frequency often precedes the onset of a calcium spike.[9,100] These observations provide valuable clues concerning the events leading up to the initiation of a regenerative calcium signal.

In the case of calcium-overloaded cardiac cells, a spark that is capable of recruiting neighboring RyRs to create a 'macrospark' seems to be responsible for initiating a calcium wave.[24] Likewise, the summation of caffeine-induced calcium sparks, i.e., the SMOCs described above, is responsible for initiating a calcium spike in sympathetic neurons.[100] Such calcium sparking may not be confined to RyRs because similar activity, referred to as calcium puffs, have been described for the $InsP_3Rs$ found in *Xenopus* oocytes.[137,190] When the level of $InsP_3$ is increased by flash photolysis of a caged precursor, localized regions in the cytoplasm release 'puffs' of calcium which spawn regenerative waves if the level of $InsP_3$ is increased. It is reasonable to speculate, therefore, that the pacemaker activity which precedes the onset of many calcium spikes represents spontaneous sparks and puffs that gradually reach the crescendo we recognize as a global calcium signal.

5.1.7. Calcium Entry

In addition to controlling the release of calcium from intracellular stores, RyRs may also regulate the influx of calcium from the extracellular medium. To date, the evidence suggesting that RyRs may control calcium influx is indirect, and relies on observations that caffeine can induce calcium entry. Four different mechanisms have been proposed to explain caffeine-dependent calcium influx (Figure 2c).

Capacitative calcium entry — Depletion of $InsP_3$-sensitive calcium stores can trigger calcium entry.[5,142] Evidence for a similar capacitative mechanism involving depletion of caffeine-sensitive stores is somewhat equivocal. For example, in porcine coronary artery smooth muscle cells, caffeine gave a transient $[Ca^{2+}]_i$ rise, without the prolonged response indicative of calcium entry. Furthermore, application of ryanodine blocked the initial response to acetylcholine, but had no effect on the subsequent prolonged response.[77] These data suggest that capacitative calcium entry in these cells is regulated primarily by $InsP_3$-sensitive calcium stores. However, capacitative calcium entry, evoked by caffeine treatment, has been described using PC12 cells[27] and guinea-pig jejunal smooth muscle cells.[135] In some cells, therefore, it appears that emptying calcium stores with caffeine may induce calcium entry through a capacitative mechanism, although the precise nature of this pathway remains to be defined (Figure 2c, i). Interestingly, caffeine-sensitive stores in smooth muscle cells appear to refill less rapidly after depletion than $InsP_3$-sensitive stores,[74] suggesting that the $InsP_3R$-linked calcium entry pathway is more effective. Additionally, caffeine-sensitive stores appear to be quite volatile, as short incubations in calcium-depleted medium inhibited a subsequent response to caffeine, although it had little effect on the response to an $InsP_3$-linked agonist.[12,102] The volatility of caffeine-sensitive stores coupled with the slow refilling time could greatly affect the possible contribution of RyRs to capacitative calcium entry.

Calcium-activated calcium entry — Some cells appear to have a nonspecific cation channel that can be activated by the elevation of cytosolic calcium (Figure 2c, ii). Mobilization of intracellular calcium stores by caffeine was found to stimulate such calcium-activated calcium entry in rat portal vein smooth muscle cells,[94] canine gastric smooth muscle,[162] and in tracheal myocytes.[72]

Plasma membrane RyR — Toad gastric smooth muscle cells display a large (80 pS) inward current that is activatable by caffeine.[54,55] The size of this current, and its sensitivity to caffeine, distinguishes it from those channels believed to be responsible for capacitative calcium entry,[36] and therefore suggests a separate mechanism. One possibility is that these smooth muscle cells have RyRs inserted into the plasma membrane, gating calcium directly into the cell (Figure 2c, iii).

Conformational coupling — Another plausible scheme, suggested to account for the caffeine-dependent calcium current in toad gastric smooth muscle, is a conformational coupling mechanism (Figure 2c, iv),[55] where a RyR in the endoplasmic reticulum makes direct contact with a calcium channel in the plasma membrane. The binding of caffeine to the RyR induces a conformational change in the protein, which is transmitted to the plasma membrane channel, causing it to open and promote calcium influx.

5.1.8. Calcium Buffering

So far, attention has focused on RyRs as release channels to generate calcium signals. If the RyR-sensitive stores are empty, they can act as a calcium sink and thus contribute to a more rapid recovery from elevated $[Ca^{2+}]_i$. Consistent with this idea, it has been demonstrated that RyR-sensitive stores may sometimes act as buffers to help curtail calcium signals in various cell types.[3,21,43,91,92,176] Such an arrangement might be a particular feature of central neurons (neocortical and hippocampal) which normally have little calcium in their caffeine-sensitive stores.[159] However, when given a brief depolarizing pulse of high potassium these stores load up, but then empty after about 10 min. Similar observations have been reported for rat cerebellar neurons.[11] On the other hand, peripheral neurons such as dorsal root ganglion cells maintain large stores of calcium. It is perhaps significant that it is these peripheral neurons that display the spontaneous sparking activity described earlier.

6. CONCLUSION

The role of RyRs in calcium signaling was first recognized in muscle cells, where they regulate contraction by releasing calcium from SR. However, accumulating evidence is suggesting that RyR expression is not restricted to excitable tissues. Initial attempts at characterizing RyR expression were pharmacologically based, relying on agents such as caffeine and ryanodine. However, these data are now being superseded by more specific molecular techniques. To date, such molecular analysis has identified a family of RyRs with differential tissue distributions, including many nonexcitable cell types.

It is also becoming increasingly apparent that RyRs contribute to cellular calcium signals when they are present. An elegant way of demonstrating the potential contribution of RyRs to calcium signals during cell stimulation is to employ the use-dependent action of ryanodine, which can lock RyRs in an open conformation. This procedure has confirmed that RyRs do indeed open during the action of many different stimuli, acting on a variety of cell types.

The next problem is to consider how such signals arriving at the cell surface can open RyRs located on the internal stores. In skeletal muscle, RyR opening depends upon conformational coupling, whereby a voltage sensor in the sarcolemma makes direct contact with RyR1. An alternative mechanism has evolved in cardiac muscle, where a small pulse of trigger calcium entering through VOCs is responsible for opening RyR2. There are suggestions that a similar 'calcium synapse' strategy may be employed in smooth muscle and neurons.

In many cell types, RyRs coexist with $InsP_3Rs$, and there is growing evidence that these two intracellular calcium release channels may interact with each other during cell stimulation. Just how these two proteins interact has proved difficult to determine, since they have very similar properties. In particular, they both display the process of CICR. This regenerative process is probably responsible for the generation of calcium spikes and waves in intact cells. RyR-dependent amplification of calcium signals originating from $InsP_3Rs$ has been demonstrated in several cell types, and it is possible that the inverse process may occur, but this presupposes that there is an intracellular messenger, like $InsP_3$, capable of activating RyR.

Such a messenger role may be performed by cyclic ADP ribose, which can mobilize calcium via RyRs in a variety of cells. However, before cyclic ADP ribose can be accepted as a calcium-mobilizing intracellular messenger, more evidence is required to show that its concentration changes during cell stimulation.

Despite the similarities between RyRs and $InsP_3Rs$, there are some important distinctions. For example, all the $InsP_3Rs$ described to date have a similar activation mechanism, i.e., $InsP_3$ binding. However RyRs display at least three fundamentally different activation mechanisms: conformational coupling, CICR, and cyclic ADP ribose binding. These alternative activation mechanisms are utilized in a tissue-specific manner, conveying calcium signals appropriate for particular cellular functions. In addition, RyR-bearing calcium stores have some properties that distinguish them from $InsP_3$-sensitive stores, such as their volatility, variable size, and the apparent lack of calcium entry following depeletion of caffeine-sensitive stores in some cell types.

Taken together, these data may indicate that RyRs are intended to play a more flexible role in calcium signaling than their $InsP_3R$ counterparts. Both proteins have similar biochemical properties, but RyRs seem to provide cells with calcium stores that have additional features and alternative activation mechanisms.

The challenge for the future is to understand further how RyRs contribute to calcium signaling, especially in those cell types where they coexist with $InsP_3Rs$, and where the activation mechanism is as yet unclear.

REFERENCES

1. **Adams, B. A., Tanabe, T., Mikami, A., Numa, S., and Beam, K. G.,** Intramembrane charge movement restored in dysgenic skeletal muscle by injection of dihydropyridine receptor cDNAs, *Nature*, 346, 569, 1990.
2. **Anderson, K., Lai, F. A., Liu, Q,-Y., Rousseau, E., Erickson, H. P., and Meissner, G.,** Structural and functional characterization of the purified cardiac ryanodine receptor-Ca^{2+} release channel complex, *J. Biol. Chem.*, 264, 1329, 1989.
3. **Barry, V. A. and Cheek, T. R.,** A caffeine- and ryanodine-sensitive intracellular Ca^{2+} store can act as a Ca^{2+} source and a Ca^{2+} sink in PC12 cells, *Biochem. J.*, 300, 589, 1994.
4. **Benham, C. D. and Bolton, T. B.,** Spontaneous transient outward currents in single visceral and vascular smooth muscle cells of the rabbit, *J. Physiol.*, 381, 385, 1986.
5. **Berridge, M. J.,** Calcium oscillations, *J. Biol. Chem.*, 265, 9583, 1990.
6. **Berridge, M. J.,** Inositol trisphosphate and calcium signaling, *Nature*, 361, 315, 1993.
7. **Bezprozvanny, I., Watras, J., and Erlich, B. E.,** Bell-shaped calcium-response curves of Ins(1,4,5)P_3- and calcium-gated channels from endoplasmic reticulum of cerebellum, *Nature*, 351, 751, 1991.

8. **Bezprozvanny, I. and Ehrlich, B. E.,** ATP modulates the function of inositol 1,4,5-trisphosphate-gated channels at two sites, *Neuron*, 10, 1175, 1993.
9. **Bolton, T. B. and Lim, S. P.,** Properties of calcium stores and transient outward currents in single smooth muscle cells of rabbit intestine, *J. Physiol.*, 409, 385, 1989.
10. **Bootman, M. D.,** Quantal Ca^{2+} release from $InsP_3$-sensitive intracellular Ca^{2+} stores, *Mol. Cell. Endocrinol.*, 98, 157, 1994.
11. **Brorson, J. R., Bleakman, D., Gibbons, S. J., and Miller, R. J.,** The properties of intracellular calcium stores in cultured rat cerebellar neurons, *J. Neurosci.*, 11, 4024, 1991.
12. **Buchan, K. W. and Martin, W.,** Bradykinin induces elevations of cytosolic calcium through mobilization of intracellular and extracellular pools in bovine aortic endothelilal cells, *Br. J. Pharmacol.*, 102, 35, 1991.
13. **Buck, E., Zimanyi, I., Abramson, J., and Pessah, I. N.,** Ryanodine stabilizes multiple conformational states of the skeletal muscle Ca^{2+} release channel, *J. Biol. Chem.*, 267, 23560, 1992.
14. **Buck, W. R., Rakow, T. L., and Shen, S. S.,** Synergistic release of calcium in sea urchin eggs by caffeine and ryanodine, *Exp. Cell Res.*, 202, 59, 1992.
15. Callewaert, G., Excitation-contraction coupling in mammalian cardiac cells, *Cardiovasc. Res.*, 26, 923, 1992.
16. **Chad, J. E. and Eckert, R.,** Calcium domains associated with individual calcium channels can account for anomalous voltage relations of Ca-dependent responses, *Biophys. J.*, 45, 993, 1984.
17. **Chadwick, C. C., Saito, A., and Fleischer, S.,** Isolation and characterization of the inositol trisphosphate receptor from smooth muscle, *Proc. Natl. Acad. Sci. U.S.A*,. 87, 2132, 1990.
18. **Cheek, T. R., Barry, V. A., Berridge, M. J. and Missiaen, L.,** Bovine adrenal chromaffin cells contain an inositol 1,4,5-trisphosphate-insensitive but caffeine-sensitive Ca^{2+} store that can be regulated by intralumenal free Ca^{2+}, *Biochem. J.*, 275, 697, 1991.
19. **Cheek, T. R., Berridge, M. J., Moreton, R. B., Stauderman, K. A., Murawsky, M. M., and Bootman, M. D.,** Quantal Ca^{2+} mobilization by ryanodine receptors is due to an all-or-none release from functionally discrete intracellular stores, *Biochem. J.*, 301, 879, 1994.
20. **Cheek, T. R., Moreton, R. B., Berridge, M. J., Stauderman, K. A., Murawsky, M. M., and Bootman, M. D.,** Quantal Ca^{2+} release from caffeine-sensitive stores in adrenal chromaffin cells, *J. Biol. Chem.*, 268, 27076, 1993.
21. **Cheek, T. R., O'Sullivan, A. J., Moreton, R. B., Berridge, M. J., and Burgoyne, R. D.,** The caffeine-sensitive Ca^{2+} store in bovine adrenal chromaffin cells; an examination of its role in triggering secretion and Ca^{2+} homeostasis, *FEBS Lett.*, 26, 91, 1990.
22. **Chen, G. and Cheung, D. W.,** Pharmacological distinction of the hyperpolarization response to caffeine and acetylcholine in guinea-pig coronary endothelial cells, *Eur. J. Pharmacol.*, 223, 33, 1992.
23. **Chen, S. R. W., Zhang, L., and MacLennan, D. H.,** Characterization of a Ca^{2+} binding and regulatory site in the Ca^{2+} release channel (ryanodine receptor) of rabbit skeletal muscle sarcoplasmic reticulum, *J. Biol. Chem.*, 267, 23318, 1992.
24. **Cheng, H., Lederer, W. J., and Cannell, M. B.,** Calcium sparks: elementary events underlying excitation-contraction coupling in heart muscle, *Science*, 262, 740, 1993.
25. **Chu, A., Duaz-Muñoz, M., Hawkes, M. J., Brush, K., and Hamilton, S. L.,** Ryanodine as a probe for the functional state of the skeletal muscle sarcoplasmic reticulum calcium release channel, *Mol. Pharmacol.*, 37, 735, 1990.
26. **Cleeman, L. and Morad, M.,** Role of Ca^{2+} channel in cardiac excitation-contraction coupling in the rat: evidence from Ca^{2+} transients and contraction, *J. Physiol.*, 432, 283, 1991.
27. **Clementi, E., Scheer, H., Zacchetti, D., Fasolato, C., Pozzan, T., and Meldolesi, J.,** Receptor-activated Ca^{2+} influx. Two independently regulated mechanisms of influx stimulation coexist in neurosecretory PC12 cells, *J. Biol. Chem.*, 267, 2164, 1992.

28. **De Smedt, H., Parys, J. B., Himpens, B., Missiaen, L., and Borghgraef, R.,** Changes in the mechanism of Ca^{2+} mobilization during differentiation of BC_3H1 muscle cells, *Biochem. J.*, 273, 219, 1991.
29. **Devine, C. E., Somlyo, A. V., and Somlyo, A. P.,** Sarcoplasmic reticulum and excitation-contraction coupling in mammalian smooth muscles, *J. Cell Biol.*, 52, 690, 1972.
30. **Diaz-Muñoz, M., Hamilton, S. L., Kaetzel, M. A., Hazarika, P., and Dedman, J. R.,** Modulation of Ca^{2+} release channel activity from sarcoplasmic reticulum by annexin VI (67-kDa Calcimedin), *J. Biol. Chem.*, 265, 15894, 1990.
31. **Dulhunty, A.,** The voltage-activation of contraction in skeletal muscle, *Prog. Biophys. Molec. Biol.*, 57, 181, 1992.
32. **Ellisman, M. H., Deerinck, T. J., Ouyang, Y., Beck, C. F., Tanksley, S. J., Walton, P. D., Airey, J. A., and Sutko, J. L.,** Identification and localization of ryanodine binding proteins in the avian central nervous system, Neuron, *5*, 135, 1990.
33. **Endo, M.,** Calcium release from sarcoplasmic reticulum, *Curr. Topics Membr. Transp.*, 25, 181, 1985.
34. **Endo, M., Tanaka, M., and Ogawa, Y.,** Calcium induced release of calcium from the sarcoplasmic reticulum of skinned muscle fibres, *Nature*, 249, 469, 1970.
35. **Fabiato, A.,** Time and calcium dependent activation and inactivation of calcium-induced release of calcium from the sarcoplasmic reticulum of a skinned canine cardiac purkinje cell, *J. Gen. Physiol.*, 85, 247, 1985.
36. **Fasolato, C., Innocenti, B., and Pozzan, T.,** Receptor-operated Ca^{2+} influx. How many mechanisms for how many channels, *Trends Pharmacol. Sci.*, 15, 77, 1994.
37. **Feng, L., Pereira, B., and Kraus-Friedmann, N.,** Different localization of inositol 1,4,5-trisphosphate and ryanodine binding sites in rat liver, *Cell Calcium*, 13, 79, 1992.
38. **Fill, M. and Coronado, R.,** Ryanodine receptor channel of sarcoplasmic reticulum, *TINS*, 11, 45, 1988.
39. **Fleischer, S. and Inui, M.,** Biochemistry and biophysics of excitation-contraction coupling, *Annu. Rev. Biophys. Biophys. Chem.*, 18, 333, 1989.
40. **Fletcher, G. H. and Chiappinelli, V. A.,** Spontaneous miniature hyperpolarizations of presynaptic nerve terminals in the chick ciliary ganglion, *Brain Res.*, 579, 165, 1992.
41. **Ford, L. E. and Podolsky, R. J.,** Regenerative calcium release within muscle cells, *Science*, 167, 58, 1970.
42. **Franzini-Armstrong, C.,** Studies of the triad. I. Structure of the junction in frog twitch fibers, *J. Cell Biol.*, 47, 488, 1970.
43. **Friel, D. D. and Tsien, R. W.,** A caffeine- and ryanodine-sensitive Ca^{2+} store in bullfrog sympathetic neurones modulates effects of Ca^{2+} entry on $[Ca^{2+}]_i$, *J. Physiol.*, 450, 217, 1992.
44. **Galione, A.,** Cyclic ADP-ribose, the ADP-ribosyl cyclase pathway and calcium signaling, *Mol. Cell. Endocrinol.* 98, 125, 1994.
45. **Galione, A., Lee, H. C., and Busa, W. B.,** Ca^{2+}-induced Ca^{2+} release in sea urchin egg homogenates: modulation by cyclic ADP-ribose, *Science*, 253, 1143, 1991.
46. **Galione, A., McDougall, A., Busa, W. B., Willmott, N., Gillot, I., and Whitaker, M.,** Redundant mechanisms of calcium-induced calcium release underlying calcium waves during fertilization of sea urchin eggs, *Science*, 261, 348, 1993.
47. **Galione, A., White, A., Willmot, N., Turner, M., Potter, B. V. L., and Watson, S. P.,** cGMP mobilizes intracellular Ca^{2+} in sea urchin eggs by stimulating cyclic ADP-ribose synthesis, *Nature,* 365, 456, 1993.
48. **Ganitkevich, V. Ya. and Isenberg, G.,** Contribution of Ca^{2+}-induced Ca^{2+} release to the $[Ca^{2+}]_i$ transients in myocytes from guinea-pig urinary bladder, *J. Physiol.*, 458, 119, 1992.
49. **Giannini, G., Clementi, E., Ceci, R., Marziali, G., and Sorrentino, V.,** Expression of a ryanodine receptor-Ca^{2+} channel that is regulated by TGF-β, *Science*, 257, 91, 1992.
50. **Gilchrist, J. S. C., Belcastro, A. N., and Katz, S.,** Intralumenal Ca^{2+} dependence of Ca^{2+} and ryanodine-mediated regulation of skeletal muscle sarcoplasmic reticulum Ca^{2+} release, *J. Biol. Chem.*, 267, 20850, 1992.

51. **Gillespie, J. I., Lynn, S., Morgan, J., Lamb, H. K., and Hawkins, A. R.,** Different isoforms of the ryanodine receptor on the sarcoplasmic reticulum of pregnant, non-pregnant and cultured human myometrial smooth muscle cells, *J. Physiol.*, abstract, 480, 123P, 1994.
52. **Graier, W. F., Simecek, S., Bowles, D. K., and Sturek, M.,** Heterogeneity of caffeine- and bradykinin-sensitive Ca^{2+} stores in vascular endothelial cells, *Biochem. J.*, 300, 637, 1994.
53. **Grégoire, G., Loirand, G., and Pacaud, P.,** Ca^{2+} and Sr^{2+} entry induced Ca^{2+} release from the intracellular Ca^{2+} store in smooth muscle cells of rat portal vein, *J. Physiol.*, 472, 483, 1993.
54. **Guerrero, A., Fay, F. S., and Singer, J. J.,** Caffeine activates a Ca^{2+} permeable, non-selective cationer channel in smooth muscle cells, *J. Gen. Physiol.*, 104, 375, 1994.
55. **Guerrero, A., Singer, J. J., and Fay, F. S.,** Simultaneous measurement of Ca^{2+} release and influx into smooth muscle cells in response to caffeine, *J. Gen. Physiol.*, 104, 395, 1994.
56. **Györke, S. and Fill, M.,** Ryanodine receptor adaptation: control mechanism of Ca^{2+}-induced Ca^{2+} release in heart, *Science*, 260, 807, 1993.
57. **Györke, S. and Palade, P.,** Role of local Ca^{2+} domains in activation of Ca^{2+}-induced Ca^{2+} release in crayfish muscle fibers, *Am. J. Physiol.*, 264, C1505, 1993.
58. **Hakamata, Y., Nakai, J., Takeshima, H., and Imoto, K.,** Primary structure and distribution of a novel ryanodine receptor/calcium release channel from rabbit brain, *FEBS Lett.*, 312, 229, 1992
59. **Henkart, M. L., Landis, D. M. D., and Reese, T. S.,** Similarity of junctions between plasma membrane and endoplasmic reticulum in muscle and neurons, *J. Cell Biol.*, 70, 338, 1976.
60. **Henzi, V. and MacDermott, A. B.,** Characteristics and function of Ca^{2+}- and inositol 1,4,5-trisphosphate-releasable stores of Ca^{2+} in neurons, *Neuroscience*, 46, 251, 1992.
61. **Hermann-Frank, A., Darling, E., and Meissner, G.,** Functional characterization of the Ca^{2+}-gated Ca^{2+} release channel of vascular smooth muscle sarcoplasmic reticulum, *Pflügers Archiv.*, 418, 353, 1991.
62. **Hernández-Cruz, A., Sala, F., and Adams, P. R.,** Subcellular calcium transients visualized by confocal microscopy in a voltage-clamped vertebrate neuron, *Science*, 247, 858, 1990.
63. **Hohenegger, M. and Suko, J.,** Phosphorylation of the purified cardiac ryanodine receptor by exogenous and endogenous protein kinases, *Biochem. J.*, 296, 303, 1993.
64. **Holliday, J., Adams, R. J., Sejnowski, T. J., and Spitzer, N. C.,** Calcium-induced release of calcium regulates differentiation of cultured spinal neurons, *Neuron*, 7, 787, 1991.
65. **Hua, S.-Y., Nohmi, M., and Kuba, K.,** Characteristics of Ca^{2+} release induced by Ca^{2+} influx in cultured bullfrog sympathetic neurones, *J. Physiol.*, 464, 245, 1993.
66. **Hua, S.-Y., Tokimasa, T., Takasawa, S., Furuyama, Y., Nohmi, M., Okamoto, H., and Kuba, K.,** Cyclic ADP-ribose modulates Ca^{2+} release channels for activation by physiological Ca^{2+} entry in bullfrog sympathetic neurons, *Neuron*, 12 1073, 1994.
67. **Iino, M., Kobayashi, T., and Endo, M.,** Use of ryanodine for functional removal of the calcium store in smooth muscle cells of the guinea pig, *Biochem. Biophys. Res. Commun.*, 152, 417, 1988.
68. **Iino, M., Yamazawa, T., Miyashita, Y., Endo, M., and Kasai, H.,** Critical intracellular Ca^{2+} concentration for all-or-none Ca^{2+} spiking in single smooth muscle cells, *EMBO J.*, 12, 5287, 1993.
69. **Ikemoto, N., Ronjat, M., Mészáros, K. G., and Koshita, M.,** Postulated role of calsequestrin in the regulation of calcium release from sarcoplasmic reticulum, *Biochemistry*, 28, 6764, 1989.
70. **Irving, A. J., Collingridge, G. L., and Schofield, J. G.,** Interactions between Ca^{2+} mobilizing mechanisms in cultured rat cerebellar granule cells, *J. Physiol.*, 456, 667, 1992.

71. **Ito, K., Ikemoto, T., and Takamura, S.,** Involvement of Ca^{2+} influx-induced Ca^{2+} release in contractions of intact vascular smooth muscle, *Am. J. Physiol.*, 261, H1464, 1991.
72. **Janssen, L. J. and Sims, S. M.,** Acetylcholine activates non-selective cation and chloride conductances in canine and guinea-pig tracheal myocytes, *J. Physiol.*, 453, 197, 1992.
73. **Kalthof, B., Bechem, M., Flocke, K., Pott, L., and Schramm, M.,** Kinetics of ATP-induced Ca^{2+} transients in cultured pig aortic smooth muscle cells depend on ATP concentration and stored Ca^{2+}, *J. Physiol.*, 466, 245, 1993.
74. **Kanaide, H., Shogakiuchi, Y., and Nakamura, M.,** The norepinephrine-sensitive Ca^{2+}-storage site differs from the caffeine-sensitive site in vascular smooth muscle of the rat aorta, *FEBS Lett.*, 214, 130, 1987.
75. **Kanmura, Y., Missiaen, L., Raeymakers, L., and Casteels, R.,** Ryanodine reduces the amount of calcium in intracellular stores of smooth muscle cells of the rabbit ear artery, *Pflügers Archiv.*, 413, 153, 1988.
76. **Kasai, H., Li, Y. X., and Miashita, Y.,** Subcellular distribution of Ca^{2+} release channels underlying Ca^{2+} waves and oscillations in exocrine pancreas, *Cell*, 74, 669, 1993.
77. **Katsuyama, H., Ito, S., Itoh, T., and Kuriyama, H.,** Effects of ryanodine on acetylcholine-induced Ca^{2+} mobilization in single smooth muscle cells of the porcine coronary artery, *Pflügers. Archiv.*, 419, 460, 1991.
78. **Kawasaki, T. and Kasai, M.,** Regulation of calcium channel in sarcoplasmic reticulum by calsequestrin, *Biophys. Biochem. Res. Commun.,* 199, 1120, 1994.
79. **Kijima, Y., Saito, A., Jetton, T. L., Magnuson, M. A., and Fleischer, S.,** Different intracellular localization of inositol 1,4,5-trisphosphate and ryanodine receptors in cardiomyocytes, *J. Biol. Chem.*, 268, 3499, 1993.
80. **Kostyuk, P. G. and Kirischuk, S. I.,** Spatial heterogeneity of caffeine- and inositol 1,4,5-trisphosphate-induced Ca^{2+} transients in isolated snail neurons, *Neuroscience*, 53, 943, 1993.
81. **Kuba, K., Morita, K., and Nohmi, M.,** Origin of calcium ions involved in the generation of a slow after hyperpolarization in bullfrog sympathetic neurones, *Pflügers. Archiv.*, 399, 194, 1983.
82. **Kuwajima, G., Futatsugi, A., Niinobe, M., Nakanishi, S., and Mikoshiba, K.,** Two types of ryanodine receptors in mouse brain: skeletal muscle type exclusively in Purkinje cells and cardiac muscle type in various neurons, *Neuron*, 9, 1133, 1992.
83. **Lai, F. A., Erickson, H. P., Rousseau, E., Liu, Q.-Y., and Meissner, G.,** Purification and reconstitution of the calcium release channel from skeletal muscle, *Nature*, 331, 315, 1988.
84. **Lea, T. J. and Ashley, C. C.,** Ca^{2+} release from the sarcoplasmic reticulum of barnacle myofibrillar bundles initiated by photolysis of caged Ca^{2+}, *J. Physiol.*, 427, 435, 1990.
85. **Lechleiter, J. C. and Clapham, D. E.,** Molecular mechanisms of intracellular calcium excitability in *Xenopus laevis* oocytes, *Cell*, 69, 283, 1992.
86. **Lee, H. C.,** Potentiation of calcium- and caffeine-induced calcium release by cyclic ADP ribose, *J. Biol. Chem.*, 268, 293, 1993.
87. **Lee, H. C., Aarhus, R., Graeff, R., Gurnack, M. E., and Walseth, T. F.,** Cyclic ADP ribose activation of the ryanodine receptor is mediated by calmodulin, *Nature*, 370, 307, 1994.
88. **Lesh, R. E., Marks, A. R., Somlyo, A. V., Fleischer, S., and Somlyo, A. P.,** Anti-ryanodine receptor antibody binding sites in vascular and endocardial endothelium, *Circ. Res.*, 72, 481, 1993.
89. **Lipp, P. and Niggli, E.,** Modulation of Ca^{2+} release in cultured neonatal rat cardiac myocytes: insight from subcellular release patterns revealed by confocal microscopy, *Circ. Res.*, 74, 979, 1994.
90. **Lipp, P., Pott, L., Callewaert, G., and Carmeliet, E.,** Calcium transients caused by calcium entry are influenced by the sarcoplasmic reticulum in guinea-pig atrial myocytes, *J. Physiol.*, 454, 321, 1992.

91. **Lipscombe, D., Madison, D. V., Poenie, M., Reuter, H., Tsien, R. W., and Tsien, R. Y.,** Imaging of cytosolic Ca^{2+} transients arising from Ca^{2+} stores and Ca^{2+} channels in sympathetic neurons, *Neuron*, 1, 355, 1988.
92. **Liu, P.-S., Lin, Y.-J., and Kao, L.-S.,** Caffeine-sensitive calcium stores in bovine adrenal chromaffin cells, *J. Neurochem.*, 56, 172, 1991.
93. **Llano, I., Di Polo, R., and Marty, A.,** Calcium-induced calcium release in cerebellar Purkinje cells, *Neuron*, 12, 663, 1994.
94. **Loirand, G., Pacaud, P., Baron, A., Mironneau, C., and Mironneau, J.,** Large conductance calcium-activated non-selective cation channel in smooth muscle cells isolated from rat portal vein, *J. Physiol.*, 437, 461, 1991.
95. **Lu, X., Xu, L., and Meissner, G.,** Activation of the skeletal muscle calcium release channel by a cytoplasmic loop of the dihydropyridine receptor, *J. Biol. Chem.*, 269, 6511, 1994.
96. **Lynn, S., Morgan, J. M., Gillespie, J. I., and Greenwell, J. R.,** A novel ryanodine sensitive calcium release mechanism in cultured human myometrial smooth muscle cells, *FEBS Lett.*, 330, 227, 1993.
97. **Ma, J., Fill, M., Knudson, C. M., Campbell, K. P., and Coronado, R.,** Ryanodine receptor of skeletal muscle is a gap junction-type channel, *Science,* 242, 99, 1988.
98. **Malgaroli, A., Fesce, R., and Meldolesi, J.,** Spontaneous $[Ca^{2+}]_i$ fluctuations in rat chromaffin cells do not require inositol 1,4,5-trisphosphate elevations but are generated by a caffeine- and ryanodine-sensitive intracellular Ca^{2+} store, *J. Biol. Chem.*, 265, 3005, 1990.
99. **Marks, A. R., Tempst, P., Hwang, K. S., Taubman, M. B., Inui, M., Chadwick, C. C., Fleischer, S., and Nadal-Ginard, B.,** Molecular cloning and characterization of the ryanodine receptor/junctional channel complex cDNA from skeletal muscle sarcoplasmic reticulum, *Proc. Natl. Acad. Sci. U.S.A.*, 86, 8683, 1989.
100. **Marrion, N. V. and Adams, P. R.,** Release of intracellular calcium and modulation of membrane currents by caffeine in bullfrog sympathetic neurons, *J. Physiol.*, 445, 515, 1992.
101. **Marty, I., Robert, M., Villaz, M., De Jong, K. S., Lai, Y., Catterall, W. A., and Ronjat, M.,** Biochemical evidence for a complex involving the dihydropyridine receptor and ryanodine receptor in triad junctions of skeletal muscle, *Proc. Natl. Acad. Sci. U.S.A.*, 91, 2270, 1994.
102. **Masuo, M., Toyo-oka, T., Shin, W. S., and Sugimoto, T.,** Growth-dependent alterations of intracellular Ca^{2+}-handling mechanisms of vascular smooth muscle cells, *Circ. Res.*, 69, 1327, 1991.
103. **Mathers, D. A. and Barker, J. L.,** Spontaneous voltage and current fluctuations in tissue cultured mouse dorsal root ganglion cells, *Brain Res.*, 293, 35, 1984.
104. **McPherson, P. S. and Campbell, K. P.,** Solubilization and biochemical characterization of the high affinity $[^3H]$ryanodine receptor from rabbit brain membranes, *J. Biol. Chem.*, 265, 18454, 1990.
105. **McPherson, P. S. and Campbell, K. P.,** The ryanodine receptor/Ca^{2+} release channel, *J. Biol. Chem.*, 268, 13765, 1993.
106. **McPherson, S. M., McPherson, P. S., Mathews, L., Campbell, K. P., and Longo, F. J.,** Cortical localization of a calcium release channel in sea urchin eggs, *J. Cell Biol.*, 116, 1111, 1992.
107. **McNulty, T. J. and Taylor, C. W.,** Caffeine-stimulated Ca^{2+} release from the intracellular stores of hepatocytes is not mediated by ryanodine receptors, *Biochem. J.*, 291, 799, 1993.
108. **Meissner, G.,** Ryanodine activation and inhibition of the Ca^{2+} release channel of sarcoplasmic reticulum, *J. Biol. Chem.*, 261, 6300, 1986.
109. **Meissner, G. and El-Hashem, A.,** Ryanodine as a functional probe of the skeletal muscle sarcoplasmic reticulum Ca^{2+} release channel, *Mol. Cell. Biochem.*, 114, 119, 1992.
110. **Mészáros, L. G., Bak, J., and Chu, A.,** Cyclic ADP-ribose as an endogenous regulator of the nonskeletal type ryanodine receptor Ca^{2+} channel, *Nature*, 364, 76, 1993.

111. **Missiaen, L., Declerck, I., Droogmans, G., Plessers, L., De Smedt, H., and Casteels, R.,** Agonist-dependent Ca^{2+} and Mn^{2+} entry dependent upon the state of filling of Ca^{2+} stores in aortic smooth muscle cells of the rat, *J. Physiol.*, 427, 171, 1990.
112. **Missiaen, L., De Smedt, H., Droogmans, G., Himpens, B., and Casteels, R.,** Calcium ion homeostasis in smooth muscle, *Pharmac. Ther.*, 56, 191, 1992.
113. **Missiaen, L., Parys, J. B., De Smedt, H., Oike, M., and Casteels, R.,** Partial calcium release in response to submaximal inositol 1,4,5-trisphosphate receptor activation, *Mol. Cell. Endocrinol.*, 98, 147, 1994.
114. **Miyazaki, S., Yazuki, M., Nakada, K., Shirakawa, H., Nakanishi, S., Nakade, S., and Mikoshiba, K.,** Block of Ca^{2+} wave and Ca^{2+} oscillation by antibody to the inositol 1,4,5-trisphosphate receptor in fertilized hamster eggs, *Science*, 257, 251, 1992.
115. **Moschella, M. C. and Marks, A. R.,** Inositol 1,4,5-trisphosphate receptor expression in cardiac myocytes, *J. Cell Biol.*, 120, 1137, 1993.
116. **Nagasaki, K. and Fleischer, S.,** Ryanodine sensitivity of the calcium release channel of sarcoplasmic reticulum, *Cell Calcium*, 9, 1, 1988.
117. **Nagasaki, K. and Kasai, M.,** Calcium-induced calcium release from sarcoplasmic reticulum vesicles, *J. Biochem.*, 90, 749, 1981.
118. **Nagasaki, K. and Kasai, M.,** Channel selectivity and gating specificity of calcium-induced calcium release channel in isolated sarcoplasmic reticulum, *J. Biochem.*, 96, 1769, 1984.
119. **Nakai, J., Imagawa, T., Hakamata, Y., Shigekawa, M., Takeshima, H., and Numa, S.,** Primary structure and functional expression from cDNA of the cardiac ryanodine receptor/ calcium release channel, *FEBS Lett.*, 271, 169, 1990.
120. **Nakanishi S., Kuwajima, G., and Mikoshiba, K.,** Immunohistochemical localization of ryanodine receptors in mouse central nervous system, *Neurosci. Res.*, 15, 130, 1992.
121. **Naruse, K. and Sokabe, M.,** Involvement of stretch-activated ion channels in Ca^{2+} mobilization to mechanical stretch in endothelial cells, *Am. J. Physiol.*, 264, C1037, 1993.
122. **Nathanson, N. M.,** Cellular and subcellular calcium signaling in gastrointestinal epithelium, *Gastroenterology*, 106, 1349, 1994.
123. **Nathanson, M. H., Padfield, P. J., O'Sullivan, A. J., Burgstahler, A. D., and Jamieson, J. D.,** Mechanism of Ca^{2+} wave propagation in pancreatic acinar cells, *J. Biol. Chem.*, 267, 18118, 1992.
124. **Neher, E. and Augustine, G. J.,** Calcium gradients and buffers in bovine chromaffin cells, *J. Physiol.*, 450, 273, 1992.
125. **Nelson, T. E. and Nelson, K. E.,** Intra- and extra lumenal sarcoplasmic reticulum membrane regulatory sites for Ca^{2+}-induced Ca^{2+} release, *FEBS Lett.*, 263, 292, 1990.
126. **Neylon, C. B. and Irvine, R. F.,** Synchronized repetitive spikes in cytoplasmic calcium in confluent monolayers of human umbilical vein endothelial cells, *FEBS Lett.*, 275, 173, 1990.
127. **Niggli, E. and Lederer, W. J.,** Voltage-independent calcium release in heart muscle, *Science*, 250, 565, 1990.
128. **Nohmi, M., Hua, S.-Y., and Kuba, K.,** Intracellular calcium dynamics in response to action potentials in bullfrog sympathetic ganglion cells, *J. Physiol.*, 458, 171, 1992.
129. **Ondrias, K., Borgatta, L., Kim, D. H., and Erlich, B. E.,** Biphasic effects of doxorubicin on the calcium release channel from sarcoplasmic reticulum of cardiac muscle, *Circ. Res.*, 67, 1167, 1990.
130. **O'Neill, S. C., Mill, J. G., and Eisner, D. A.,** Local activation of contraction in isolated rat ventricular myocytes, *Am. J. Physiol.*, 258, C1165, 1990.
131. **Osipchuk, Y. V., Wakui, M., Yule, D. I., Gallacher, D. V., and Petersen, O. H.,** Cytoplasmic Ca^{2+} oscillations evoked by receptor stimulation, G-protein activation, internal application of inositol trisphosphate or Ca^{2+}. Simultaneous microfluorimetry and Ca^{2+}-dependent Cl^- current recording in single pancreatic acinar cells, *EMBO J.*, 9, 697, 1990.

132. **Otsu, K., Willard, H. F., Khanna, V. K., Zorzato, F., Green, N. M., and MacLennan, D. H.,** Molecular cloning of cDNA encoding the Ca^{2+} release channel (ryanodine receptor) of rabbit cardiac muscle sarcoplasmic reticulum, *J. Biol. Chem.*, 265, 13472, 1990.
133. **Ouyang, Y., Deerinck, T. J., Walton, P. D., Airey, J. A., Sutko, J. L. and Ellisman, M. H.,** Distribution of ryanodine receptors in the chicken central nervous system, *Brain Res.*, 620, 269, 1993.
134. **Oyamada, H., Iino, M., and Endo, M.,** Effects of ryanodine on the properties of Ca^{2+} release from the sarcoplasmic reticulum in skinned skeletal muscle fibers of the frog, *J. Physiol.*, 470, 335, 1993.
135. **Pacaud, P. and Bolton, T. B.,** Relation between muscarinic receptor cationic current and internal calcium in guinea-pig jejunal smooth muscle cells, *J. Physiol.*, 441, 477, 1991.
136. **Padua, R. A., Nagy, J. I., and Geiger, J. D.,** Ionic strength dependence of calcium, adenine nucleotide, magnesium and caffeine actions on ryanodine receptors in rat brain, *J. Neurochem.*, 62, 2340, 1994.
137. **Parker, I. and Yao, Y.,** Regenerative release of calcium from functionally discrete subcellular stores by inositol trisphosphate, *Proc. R. Soc. Lond. B.*, 246, 269, 1991.
138. **Parys, J. B., Sernett, S. W., DeLisle, S., Snyder, P. M., Welsh, M. J., and Campbell, K. P.,** Isolation and characterization of the inositol 1,4,5-trisphosphate receptor protein in *Xenopus laevis* oocytes, *J. Biol. Chem.*, 267, 18776, 1992.
139. **Penner, R., Neher, E., Takeshima, H., Nishimura, S., and Numa, S.,** Functional expression of the calcium release channel from skeletal muscle ryanodine receptor cDNA, *FEBS Lett.*, 259, 217, 1989.
140. **Pessah, I. N., Stambuck, R. A., and Cassida, J. E.,** Ca^{2+}-activated ryanodine binding: mechanisms of sensitivity and intensity modulation by Mg^{2+}, caffeine and adenine nucleotides, *Mol. Pharmacol.*, 31, 232, 1987.
141. **Pessah, I. N. and Zimanyi, I.,** Characterization of multiple [^{3}H]ryanodine binding sites on the Ca^{2+} release channel of sarcoplasmic reticulum from skeletal and cardiac muscle: evidence for a sequential mechanism in ryanodine action, Mol. Pharmacol., 39, 679, 1991.
142. **Putney, J. W., Jr.,** Capacitative calcium entry revisited, *Cell Calcium*, 11, 611, 1990.
143. **Randriamampita, C., Bismuth, G., and Trautman, A.,** Ca^{2+}-induced Ca^{2+} release amplifies the Ca^{2+} response elicited by inositol trisphosphate in macrophages, *Cell Reg.*, 2, 513-522, 1991.
144. **Reber, B. F. X., Stucki, J. W., and Reuter, H.,** Unidirectional interaction between two intracellular calcium stores in rat phaeochromocytoma (PC12) cells, *J. Physiol.*, 468, 711, 1993.
145. **Rios, E., Pizarro, G., and Stefani, E.,** Charge movement and the nature of signal transduction in skeletal muscle excitation-contraction coupling, *Annu. Rev. Physiol.*, 54, 109, 1992.
146. **Roe, M. W., Lancaster, M. E., Mertz, R. J., Worley, J. F., III, and Dukes, I. D.,** Voltage-dependent intracellular calcium release from mouse islets stimulated by glucose, *J. Biol. Chem.*, 268, 9953, 1993.
147. **Rousseau, E., Smith, J. S., and Meissner, G.,** Ryanodine modifies conductance and gating behavior of single Ca^{2+} release channels, *Am. J. Physiol.*, 253, C364, 1987.
148. **Rousseau, E., LaDine, J., Liu, Q.-Y., and Meissner, G.,** Activation of the Ca^{2+} release channel of skeletal muscle sarcoplasmic reticulum by caffeine and related compounds, *Archiv. Biochem. Biophys.*, 267, 75, 1988.
149. **Saida, K. and van Breemen, C.,** A possible Ca^{2+}-induced Ca^{2+} release mechanism mediated by norepinephrine in vascular smooth muscle, *Pflügers Archiv.*, 397, 166, 1983.
150. **Sah, P., Dulhunty, A., Junakar, P., and Stanhope, C.,** Subcellular distribution of ryanodine receptor-like and calcium ATPase-like immunoreactivity in brainstem and cerebellar neurons of rat and guinea pig, *Neurosci. Lett.*, 166, 143, 1994.

151. **Sah, P. and McLachlan, E. M.,** Ca^{2+}-activated K^+ current underlying the after hyperpolarization in guinea pig vagal neurons: a role for Ca^{2+}-activated Ca^{2+} release, *Neuron*, 7, 257, 1991.
152. **Sanchez-Bueno, A. and Cobbold, P. H.,** Agonist-specificity in the role of Ca^{2+}-induced Ca^{2+} release in hepatocyte Ca^{2+} oscillations, *Biochem. J.*, 291, 169, 1993.
153. **Sanchez-Bueno, A., Marrero, I., and Cobbold, P. H.,** Caffeine inhibits agonists-induced cytoplasmic Ca^{2+} oscillations in single rat hepatocytes, *Biochem. Biophys. Res. Commun.*, 198, 728, 1994.
154. **Satin, L. S. and Adams, P. R.,** Spontaneous miniature outward currents in cultured bullfrog neurons, *Brain Res.*, 401, 331, 1987.
155. **Schmid, A., Dehlinger-Kremer, M., Schulz, I., and Gögelein, H.,** Voltage-dependent $InsP_3$-insensitive calcium channels in membranes of pancreatic endoplasmic reticulum vesicles, *Nature*, 346, 374, 1990.
156. **Schneider, M. F. and Simon, B. J.,** Inactivation of calcium release from sarcoplasmic reticulum in frog skeletal muscle, *J. Physiol.*, 405, 727, 1988.
157. **Sharp, A. H., McPherson, P. S., Dawson, T. M., Aoki, C., Campbell, K. P., and Snyder, S. H.,** Differential immunohistochemical localization of inositol 1,4,5-trisphosphate- and ryanodine-sensitive Ca^{2+} release channels in rat brain, *J. Neurosci.*, 13, 3051, 1993.
158. **Shirasaki, T., Harata, N., and Akaike, N.,** Metabotropic glutamate response in acutely dissociated hippocampal CA1 pyramidal neurones of the rat, *J. Physiol.*, 475, 439, 1994.
159. **Shmigol, A., Kirischuk, S., Kostyuk, P., and Verkhratsky, A.,** Different properties of caffeine-sensitive Ca^{2+} stores in peripheral and central mammalian neurones, *Pflügers. Archiv.*, 426, 174, 1994.
160. **Shoshan-Barmatz, V., Zhang, G. H., Garretson, L., and Kraus-Friedmann, N.,** Distinct ryanodine- and inositol 1,4,5-trisphosphate-binding sites in hepatic microsomes, *Biochem. J.*, 268, 699, 1990.
161. **Shoshan-Barmatz, V., Pressley, T. A., Higham, S., and Kraus-Friedmann, N.,** Characterization of high-affinity ryanodine-binding sites of rat liver endoplasmic reticulum, *Biochem. J.*, 276, 41, 1991.
162. **Sims, S. M.,** Cholinergic activation of a non-selective cation current in canine gastric smooth muscle is associated with contraction, *J. Physiol.*, 449, 377, 1992.
163. **Sitsapesan, R., McGarry, S. J., and Williams, A. J.,** Cyclic ADP-ribose competes with ATP for the adenine nucleotide binding site on the cardiac ryanodine receptor Ca^{2+}-release channel, *Circ. Res.*, 75, 596, 1994.
164. **Sitsapesan, R. and Williams, A. J.,** Regulation of the gating of the sheep cardiac sarcoplasmic reticulum Ca^{2+}-release channel by lumenal Ca^{2+}, *J. Membr. Biol.*, 137, 215, 1994.
165. **Sorrentino, V. and Volpe, P.,** Ryanodine receptors: how many, where and why?, *Trends Pharmacol. Sci.*, 14, 98, 1993.
166. **Stauderman, K. A. and Murawsky, M. M.,** The inositol 1,4,5-trisphosphate-forming agonist histamine activates a ryanodine-sensitive Ca^{2+} release mechanism in bovine adrenal chromaffin cells, *J. Biol. Chem.*, 266, 19140, 1991.
167. **Stern, M. D.,** Buffering of calcium in the vicinity of a channel pore, *Cell Calcium*, 13, 183, 1992.
168. **Stern, M. D.,** Theory of excitation-contraction coupling in cardiac muscle, *Biophys. J.*, 63, 497, 1992.
169. **Swann, K.,** Different triggers for calcium oscillations in mouse eggs involve a ryanodine-sensitive store, *Biochem. J.*, 287, 79, 1992.
170. **Takasawa, S., Nata, K., Yonekura, H., and Okamaoyo, H.,** Cyclic ADP-ribose in insulin secretion from pancreatic β cells, *Science*, 259, 370, 1993.
171. **Takata, M., Sabe, H., Hata, A., Inazu, T., Homma, Y., Nukada, T., Ymamura, H., and Kurosaki, T.,** Tyrosine kinases Lyn and Syk regulate B cell receptor-coupled Ca^{2+} mobilization through distinct pathways, *EMBO J.*, 13, 1341, 1994.

172. **Takeshima, H., Nishimura, S., Matsumoto, T., Ishida, H., Kangawa, K., Minamino, N., Matsuo, H., Ueda, M., Hanaoka, M., Hirose, T., and Numa, S.,** Primary structure and expression from complementary DNA of skeletal muscle ryanodine receptor, *Nature*, 339, 439, 1989.
173. **Tanabe, T., Beam, K. G., Adams, B. A., Niidme, T., and Numa, S.,** Regions of the skeletal muscle dihydropyridine receptor critical for excitation-contraction coupling, *Nature*, 346, 567, 1990.
174. **Taylor, C. W. and Richardson, A.,** Structure and function of inositol 1,4,5-trisphosphate receptors, *Pharmac. Ther.*, 51, 97, 1991.
175. **Thayer, S. A., Hirning, L. D., and Miller, R. J.,** The role of caffeine-sensitive calcium stores in the regulation of the intracellular free calcium concentration in rat sympathetic neurons *in vivo*, *Mol. Pharmacol.*, 34, 664, 1988.
176. **Thayer, S. A., Perney, T. M., and Miller, R. J.,** Regulation of calcium homeostasis in sensory neurons by bradykinin, *J. Neurosci.*, 8, 4089, 1988.
177. **Thorn, P., Gerasimenko, O., and Petersen, O. H.,** Cyclic ADP-ribose regulation of ryanodine receptors in agonist evoked cytosolic Ca^{2+} oscillations in pancreatic acinar cells, *EMBO J.*, 13, 2038, 1994.
178. **Vites, A. and Pappano, A.,** Inositol 1,4,5-trisphosphate releases intracellular Ca^{2+} in permeabilized chick atria, *Am. J. Physiol.*, 258, H1745, 1990.
179. **Wagner, K. A., Yacono, P. W., Golan, D. E., and Tashjian, A. H.,** Mechanism of spontaneous intracellular calcium fluctuations in single GH_4C_1 rat pituitary cells, *Biochem. J.*, 292, 175, 1993.
180. **Wakui, M., Osipchuk, Y. V., and Petersen, O. H.,** Receptor-activated cytoplasmic Ca^{2+} spiking mediated by inositol trisphosphate is due to Ca^{2+}-induced Ca^{2+} release, *Cell*, 63, 1025, 1990.
181. **Walseth, T. F., Aarhus, R., Kerr, J. A., and Lee, H. C.,** Identification of cyclic ADP-ribose-binding proteins by photoaffinity labeling, *J. Biol. Chem.*, 268, 26686, 1993.
182. **Walton, P. D., Airey, J. A., Sutko, J. L., Beck, C. F., Mignery, G. A., Südhoff, T. C., Deerinck, T. J., and Ellisman, M. H.,** Ryanodine and inositol trisphosphate receptors coexist in avian cerebellar Purkinje neurons, *J. Cell Biol.*, 113, 1145, 1991.
183. **Wang, J. and Best, P. M.,** Inactivation of the sarcoplasmic reticulum calcium channel by protein kinase, *Nature*, 359, 739, 1992.
184. **Weir, W. G.,** $[Ca^{2+}]_i$ waves in heart cells: more than a passing fancy, *Biophys. J.*, 65, 2270, 1993.
185. **Weir, W. G., Cannell, M. B., Berlin, J. R., Marban, E., and Lederer, W. J.,** Cellular and subcellular heterogeneity of $[Ca^{2+}]_i$ in single heart cells revealed by fura-2, *Science*, 235, 325, 1987.
186. **Weir, W. G., Egan, T. M., López-López, J. R., and Balke, C. W.,** Local control of excitation-contraction coupling in rat heart cells, *J. Physiol.*, 474, 463, 1994.
187. **Wibo, M. and Godfraind, T.,** Comparitive localization of inositol 1,4,5-trisphosphate and ryanodine receptors in intestinal smooth muscle: an analytical subfractionation study, *Biochem. J.*, 297, 415, 1994.
188. **Witcher, D. R., Kovacs, R. J., Schulman, H., Cefali, D. C., and Jones, L. R.,** Unique phosphorylation site on the cardiac ryanodine receptor regulates calcium channel activity, *J. Biol. Chem.*, 266, 11144, 1991.
189. **Xu, L., Lai, F. A., Cohn, A., Etter, E., Guerrero, A., Fay, F. S., and Meissner, G.,** Evidence for a Ca^{2+}-gated ryanodine-sensitive Ca^{2+} release channel in visceral smooth muscle, *Proc. Natl. Acad. Sci. U.S.A.*, 91, 3294, 1994.
190. **Yao, Y. and Parker, I.,** Ca^{2+} influx modulation of temporal and spatial patterns of inositol trisphosphate-mediated Ca^{2+} liberation in *Xenopus* oocytes, *J. Physiol.*, 476, 17, 1994.
191. **Yoshida, A., Ogura, A., Imagawa, T., Shigekawa, M., and Takahashi, M.,** Cyclic AMP-dependent phosphorylation of the rat brain ryanodine receptor, *J. Neurosci.*, 12, 1094, 1992.

192. **Zacchetti, D., Clementi, E., Fasolato, C., Lorenzon, P., Zottini, M., Grohovaz, F., Fumagalli, G., Pozzan, T., and Meldolesi, J.,** Intracellular Ca^{2+} pools in PC12 cells, *J. Biol. Chem.*, 266, 20152, 1991.
193. **Zhang, Z.-D., Kwan, C.-Y., and Daniel, E. E.,** Subcellular-membrane characterization of [^{3}H]-ryanodine-binding sites in smooth muscle, *Biochem. J.*, 290, 259, 1993.
194. **Ziegelstein, R. C., Spurgeon, H. A., Pili, R., Passaniti, A., Cheng, L., Corda, S., Lakatta, E. G., and Capogrossi, M. C.,** A functional ryanodine-sensitive intracellular Ca^{2+} store is present in vascular endothelial cells, *Circ. Res.*, 74, 151, 1994.
195. **Zorzato, F., Fujii, J., Otsu, K., Phillips, M., Green, N. M., Lai, F. A., Meissner, G., and MacLennan, D. H.,** Molecular cloning of cDNA encoding human and rabbit forms of the Ca^{2+} release channel (ryanodine receptor) of skeletal muscle sarcoplasmic reticulum, *J. Biol. Chem.*, 265, 2244, 1990.

Chapter 9

THE ROLE OF THE SKELETAL MUSCLE RYANODINE RECEPTOR GENE IN MALIGNANT HYPERTHERMIA AND CENTRAL CORE DISEASE

David H. MacLennan and Michael S. Phillips

TABLE OF CONTENTS

1. HUMAN MALIGNANT HYPERTHERMIA

Malignant hyperthermia is a clinical syndrome in which genetically susceptible individuals respond to the administration of potent inhalational anesthetics and depolarizing skeletal muscle relaxants with high fever and skeletal muscle rigidity, hyperventilation, hypoxia, and lactic acidosis.[1] Cellular damage, which brings about electrolyte imbalance and elevation in the serum and urine levels of muscle enzymes and myoglobin, can lead to

0-8493-8543-1/95/$0.00+$.50

tachycardia, arrhythmia, and unstable blood pressure. If therapy is not initiated immediately, the patient may die within minutes from ventricular fibrillation, within hours from pulmonary edema or coagulapathy, or within days from neurological damage or obstructive renal failure, resulting largely from the release of muscle proteins into the circulation.

In a modern operating room setting, anesthesiologists identify many patients at risk through case histories which include information on their kinship to individuals who have had an MH reaction. For these patients, anesthetic routines are varied to include nontriggering anesthetics. As a standard preventive step, heart rate, body temperature, and end tidal CO_2 production are monitored during the course of anesthesia. An increase in any of these may lead to the diagnosis of MH, in which case the administration of the MH-triggering anesthetic is stopped, the patient is hyperventilated with 100% oxygen, and the clinical antidote, dantrolene, is administered. These practices have lowered the death rate from MH episodes from over 80% to less than 7% in recent years, but neurological or kidney damage still contributes to the morbidity resulting from MH episodes.[1]

Human MH has been recognized as a genetic abnormality since the early 1960s.[2,3] The abnormality is inherited as an autosomal dominant mutation with incomplete penetrance. MH reactions do not occur in every case of administration of anesthetics to susceptible individuals and MH reactions occur frequently in susceptible individuals who have previously undergone uneventful general anesthetics.[1] Because of the incomplete penetrance of the gene, the difficulty in defining mild reactions and the caution and care now taken by anesthesiologists, it is difficult to determine the actual incidence of MH susceptibility genes in the general population. Britt and Kalow[4] estimated the incidence to be about 1 in 15,000 anesthetics in children and about 1 in 50,000 to 1 in 100,000 anesthetics in adults. Ording[5] found the incidence of MH episodes to be about 1 in 65,000 anesthetics in Denmark. For the reasons outlined above, these are probably underestimates of the true incidence of MH susceptibility.

1.1. DIAGNOSTIC TESTS FOR MALIGNANT HYPERTHERMIA

Most individuals who inherit an MH susceptibility gene appear healthy. However, MH episodes may occur in individuals who have inherited other muscle diseases with more deleterious phenotypes such as central core disease,[6] King-Denborough syndrome,[7,8] Duchenne muscular dystrophy,[9-12] and possibly other myopathies.[13-15] Accordingly, the most important goals of MH research are to provide therapy during an MH episode and to prevent MH episodes through identification of MH-susceptible individuals in advance of anesthesia. If MH susceptibility is known, the use of alternate anesthetics and nondepolarizing muscle relaxants can circumvent the triggering of an MH reaction.

The *in vitro* halothane caffeine contracture test (CHCT) was developed in the early 1970s on the premise that the muscle from MH-susceptible (MHS) individuals might be more sensitive to agents inducing contraction and, therefore, might contract in the presence of lower amounts of either caffeine[16] or halothane[17] than the muscle from normal individuals. The North American test protocol[18] and the European test protocol[19] have been standardized on the basis of this premise.

In the North American test, a muscle fascicle is secured at one end and immersed in Krebs-Ringer buffer at 37°C and pH 7.4. The other end of the muscle is then attached to a force displacement transducer and isometric tension is recorded with a polygraph beginning with a resting tension of 1 to 2 g. The muscle is stimulated every 5 s to assure its viability. After stable twitch and baseline tensions are achieved, caffeine is added directly to the bath in incremental doses from 0.5 to 32 m*M*. Measurement of the contracture is made 4 min after the addition of caffeine, expressed as grams of tension. The most sensitive and specific measurement is of grams tension induced at 2 m*M* caffeine. MH susceptibility is associated with an increase in tension of 0.2 g or more at 2 m*M* caffeine. An alternative measurement is of the dose of caffeine required to raise the resting tension by 1 g, termed the caffeine specific concentration (CSC). If 1 g of increased tension is achieved with 4 m*M* caffeine or less, the patient is considered to be MHS. The amplitude of contractions induced by 3% halothane is measured in separate muscle strips placed in separate baths. MHS individuals are defined as those with muscle strips which produce greater than 0.2 to 0.7 g of tension in response to 3% halothane, depending on the laboratory.

In the European protocol, halothane is added to the solution bathing a single muscle fascicle at 0.5, 1.0, 1.5, 2.0, and 3.0% by volume. A threshold of 0.2 g contracture seen at 2% (0.44 m*M*) halothane or less is considered to indicate MH susceptibility. Caffeine is also added to a second muscle fascicle at concentrations of 0.5, 1.0, 1.5, 2.0, 3.0, and 4.0 m*M* or until a threshold of 0.2 g contracture is obtained. A threshold of contracture tension of 0.2 g or more, observed at 2 m*M* caffeine or less, is considered to indicate MH susceptibility. In the European protocol, only those whose muscles respond abnormally to both caffeine and halothane are considered MHS, while those who react abnormally with one, but not the other, are considered to be MH equivocal (MHE).

The CHCT is a valuable clinical test.[20] Since failure to detect MH susceptibility can result in a serious or fatal outcome, sensitivity approaching 100% is more important for a clinical diagnosis than is specificity. The North American CHCT currently achieves 92 to 95% sensitivity and 75 to 53% specificity for the same tests.[21] As a clinical test, the CHCT assures that appropriate anesthetics are administered to all those patients who are MHS, while those diagnosed as normal can be treated with normal anesthetic routines in a cost-effective manner. Problems with the CHCT are that it is invasive and expensive to perform and, as defined above under specificity, has a significant false positive error rate[22] and a very small false negative error rate.[23] While the error rate does not pose any significant problem for clinicians, inaccurate diagnosis creates difficulties for geneticists attempting to link the inheritance of MH susceptibility to inheritance of a specific allele.

An example of the problems of using the North American test response as the basis for MH status was presented in studies in which DNA linkage tests were compared with CHCT.[24] In a Canadian family many members were diagnosed as MHS and no linkage to a potential causal gene, *RYR1*, could be established. If the diagnostic cutoff points were altered, linkage of MH susceptibility to the *RYR1* gene with a lod score (log of the odds favoring linkage) of 3.6 at a recombination frequency of 0.0 was established for this family. Because of limitations in accuracy of using the CHCT for MH diagnosis, researchers at the North American Malignant Hyperthermia Registry have spent several years clinically defining the MHS population[25]and then optimizing the North American CHCT protocols.[21,22] These studies are helping to resolve the problems concerning differentiation of MHS and normal individuals in the population.

Other *in vitro* tests for MH have been developed. These include measurement of differential contracture response to the administration of ryanodine;[26,27] measurement of cytosolic free Ca^{2+} concentration in lymphocytes, both before and after addition of halothane;[28,29] measurement of the level of resting Ca^{2+} in the muscle of MH individuals by the direct use of implanted Ca^{2+} electrodes;[30] and the use of Fura 2 to measure intracellular Ca^{2+} concentration.[31] Phosphorus nuclear magnetic resonance spectroscopy has also been tested as a means of discriminating between MHS and MHN individuals.[32] In this test, changes in ATP, phosphocreatine, inorganic phosphate, and acidity are measured noninvasively in whole muscles. None of these tests has yet proven to be sufficiently specific and sensitive to supplant the caffeine/halothane contracture test.

2. PORCINE MALIGNANT HYPERTHERMIA

Individuals among herds of lean, heavily muscled swine are frequently susceptible to fatal episodes of fever, skeletal muscle rigidity, and hyperventilation brought on by various forms of stress, including overheating, exercise, mating, or transportation to market. This syndrome is referred to as the porcine stress syndrome or PSS. When administered halothane and succinylcholine, these pigs are frequently found to be susceptible to MH reactions[33,34] and to display the same symptoms of high fever, skeletal muscle rigidity, cyanosis, hyperventilation, hypoxia, lactic acidosis, and death that characterize human MH. Stress-induced deaths occur predominantly with homozygous MH animals and rarely in heterozygotes, illustrating the recessive nature of the genetics of stress-induced deaths. Nevertheless, the heterozygous MH genotype does lead to enhanced meat production,[35-37] to intermediate sensitivity of the Ca^{2+} release channel to ligand-induced gating,[38] and to intermediate contracture responses to halothane or to halothane plus succinylcholine.[39-42] The genetic background in which the porcine MH gene is expressed, however, might influence the heterozygote response.[43]

A serious problem in the pork industry is the incidence of pale, soft, exudative (PSE) meat that is found in large segments of the carcasses of many animals.[44] A high incidence of PSE meat has been associated with a high incidence of the PSS gene. When it became possible to identify MH heterozygotes and homozygotes through a blood test,[45] it was possible to relate the incidence of PSE to the presence of the MH gene defect.[46-48] In an initial study, random samples of Canadian swine showed a 15% incidence of heterozygotes and a 0.6% incidence of homozygotes. In a second study, 913 loins were selected on the cutting line, characterized for PSE, and genotyped. MH heterozygotes (N/n genotype) accounted for 11% of the loins, while MH homozygotes (n/n genotype) accounted for 0.6%. By comparison, 28.7% of PSE loins were from N/n pigs and 3.6% were from n/n pigs. Since 67.7% of the PSE meat came from normal (N/N) animals, it is apparent that elimination of the PSS gene alone will not eradicate the PSE problem. The proportion of PSE meat is also determined by preslaughter management of pigs. The quality of meat obtained from heterozygous animals will benefit the most from improved preslaughter management practices.[48]

When the MH (PSS) gene was recognized as contributing to economic loss, both through mortality and through devalued meat products, attempts were made to remove the gene from swine populations. This turned out to be very difficult because there was no satisfactory way of identifying heterozygous carriers of the MH gene, although they could be identified retrospectively from genetic trials. Moreover, the MH gene contributes to lean body mass[35-37] and, in selecting breeding stock for such characteristics as large ham conformation, large loin eye area, and excessive leanness, breeders also select for the MH gene. The advantage of the MH gene for lean meat production was sufficiently obvious to assure its dissemination throughout the world among lean, heavily muscled breeds of swine. As a result, the incidence of the MH gene was stabilized in most lean, heavily muscled breeds of swine.

3. PHYSIOLOGICAL AND GENETIC BASIS FOR MALIGNANT HYPERTHERMIA

3.1. THE Ca^{2+} RELEASE CHANNEL

Studies from the 1950s onward of the mechanisms controlling muscle contraction revealed that the interaction of actin and myosin was regulated by Ca^{2+} and that Ca^{2+} regulation was mediated through the Ca^{2+} binding protein, troponin.[49-50] Moreover, muscle Ca^{2+} concentrations were shown to be regulated by a muscle membrane system referred to as the sarcoplasmic reticulum. Ca^{2+} was also shown to play a role in the control of glycolysis in muscle through its activation of phosphorylase kinase.[51] From such studies, it became apparent that a defect in Ca^{2+} regulation, leading to the continued presence of

Ca^{2+} within the sarcoplasm, could induce muscle contracture, extensive glycolysis, and enhanced mitochondrial oxidation of glycolytic end products. These reactions, leading to high turnover of ATP, could be responsible for the elevated temperatures associated with MH episodes.[52,53]

Studies of Ca^{2+} release channels showed that they are localized in heavy sarcoplasmic reticulum vesicles representing the terminal cisternae of the membrane system.[54,55] The development of rapid filtration assays to measure the kinetics of Ca^{2+} release from heavy terminal cisternae vesicles permitted evaluation of the regulatory effects of specific physiological and pharmacological ligands.[56-59] More precise characterization was possible when single Ca^{2+} release channel proteins were incorporated into planar lipid bilayers, where they were shown to form ligand-gated channels with a conductance greater than 100 pS in 50 m*M* Ca^{2+}.[60] Ca^{2+} release is activated by micromolar Ca^{2+}, millimolar ATP, and millimolar caffeine, and inhibited by millimolar Mg^{2+} [53,56,59-62] and micromolar calmodulin.[57]

Ca^{2+} release channels have been cloned from skeletal (*RYR1*),[63-65] cardiac (*RYR2*),[66,67] and nonmuscle sources (*RYR3*).[68,69] The proteins contain from 4872 to 5037 amino acids with masses between 550,000 and 564,000 Da. The number of transmembrane sequences near the COOH-terminus in ryanodine receptors is controversial.[63,64] Each of the three ryanodine receptor isoforms has four strongly hydrophobic sequences which are well conserved and are almost certainly transmembrane, but additional transmembrane loops may exist.

The proteins have four repeated sequences in two tandem pairs that are well conserved, but no function has been assigned to these repeat sequences. Residues 2809 and 2843 in cardiac and skeletal isoforms, respectively, are the major phosphorylation sites in these proteins.[70,71] A predicted ATP binding sequence begins at residue 2652, and the region between residues 2800 and 3050 contains predicted calmodulin binding sites. Mutations of residues 2434 and 2433 alter regulatory properties of the ryanodine receptor, giving rise to central core disease[72] and MH.[73,74] Thus, the sequence between residues 2600 and 3050 appears to be a regulatory domain.[66,72] ATP, Ca^{2+}, and calmodulin binding sites were proposed by Takeshima *et al.*[63] to cluster around the proposed transmembrane sequences in the COOH-terminal end of the protein and to encompass the region between residues 4253 to 4499, forming a second predicted modulator binding region.

As described in other chapters in this book, the regulatory cascade for Ca^{2+} release into skeletal muscle begins with depolarization of nerve and muscle membranes. Depolarization courses through the transverse tubule, an invagination of the plasma membrane into the interior of skeletal muscle cells. Polarization and depolarization are not properties of the sarcoplasmic reticulum membrane, since it is relatively permeable to monovalent ions.[75] Depolarization of the transverse tubular membrane is followed by release of Ca^{2+} from the sarcoplasmic reticulum through a Ca^{2+} release channel located at the junctional face of the sarcoplasmic reticulum,[55] the region of the sarcoplasmic reticulum that abuts and associates with the transverse tubule. During depolarization, movement of a fixed charge in the transverse tubular membrane has been observed to precede Ca^{2+} release[76] and has been proposed to be a prerequisite for Ca^{2+} release.[77] Since dihydropyridine blocks both charge movement and Ca^{2+} release, the fixed charge movement has been proposed to occur within the dihydropyridine (DHP) receptor, a slow Ca^{2+} channel which is concentrated in transverse tubular membranes.

Ca^{2+} released to the sarcoplasm is then transported back to the lumen of the sarcoplasmic reticulum by a Ca^{2+} pump located throughout all regions of the sarcoplasmic reticulum, with the exception of the junctional face membrane. Ca^{2+} is stored in association with an acidic, lumenal calcium binding protein, calsequestrin,[78] localized in the junctional terminal cisternae.[79] Its localization may be the result of an initial nucleating event brought about by its association with a basic transmembrane protein, triadin,[80-82] located exclusively at the junctional face of the sarcoplasmic reticulum.

The triad junction is made up by the close apposition of channels originating in the sarcoplasmic reticulum, with other proteins originating in the transverse tubule.[83] Morphological studies suggest that the protein in the transverse tubule that interacts with the ryanodine receptor is the dihydropyridine receptor. Block et al.[84] showed that the DHP receptors are positioned in the transverse tubule membrane in a pattern that would permit their physical interaction with alternate ryanodine receptors. A direct physical interaction between the DHP and ryanodine receptor proteins has not yet been demonstrated, but studies utilizing chimeric molecules between the cardiac and skeletal muscle α-1 subunits of the DHP receptor suggest that there is specific functional interaction between DHP and ryanodine receptors.[85] Moreover, functional interaction between ryanodine receptors incorporated into planar lipid bilayers and expressed fragments of the α-1 subunits of cardiac and skeletal muscle DHP receptors has been demonstrated.[86]

3.2. ROLE OF THE Ca^{2+} RELEASE CHANNEL IN MALIGNANT HYPERTHERMIA

Since abnormalities in regulation of the intracellular concentrations of Ca^{2+} might lead to MH, mutations in genes encoding the Ca^{2+} pump, the Ca^{2+} release channel, or other proteins that participate in the cascade of excitation-contraction coupling might be causal of MH. Abnormalities in the Ca^{2+} pump were ruled out in biochemical studies.[87-89] However, higher rates of Ca^{2+}-induced Ca^{2+} release, particularly at low levels of inducing Ca^{2+}, have been observed in membrane vesicle preparations from both human[90] and porcine[91-93] muscle, and closure of single porcine MH channels at high Ca^{2+} concentrations has been shown to be inhibited.[94,95] In comparable studies of human muscle, Ca^{2+} release channels with abnormally greater caffeine sensitivity were detected in MH individuals.[96] In sarcoplasmic reticulum from swine with MH, ryanodine binding, which is dependent on the open state of the Ca^{2+} release channel, is enhanced.[97] Digestion with trypsin revealed an alteration in the amino acid sequence of the Ca^{2+} release channel in MH animals.[98] Thus, comparative biochemical and physiological studies implicated the Ca^{2+} release channel as a potential causal factor for MH.

A defect in the Ca^{2+} release channel giving rise to abnormal Ca^{2+} regulation within skeletal muscle could account for all of the symptoms of MH (Figure 1). If the Ca^{2+} release channels had longer open times in the presence of anesthetic agents, intracellular Ca^{2+} might be chronically elevated, resulting in muscle contracture and activation of the first steps in glycogenolysis through activation of phosphorylase kinase.[51] Muscle contracture and the pumping of cytoplasmic Ca^{2+} to the lumen of the sarcoplasmic reticulum would consume large amounts of ATP, generating heat. The ADP formed would stimulate glycolysis and the mitochondrial oxidation of pyruvate derived from glucose. These hypermetabolic responses would lead to depletion of ATP, glycogen, and oxygen, to the production of excess lactic acid, CO_2, and heat and, ultimately, to the disruption of cellular and extracellular ion balance.

3.3. LINKAGE OF THE *RYR1* GENE TO MALIGNANT HYPERTHERMIA

In early studies of porcine MH, Andersen and Jensen[99] demonstrated linkage between inheritance of MH and polymorphisms in the gene encoding glucose phosphate isomerase (*GPI*). Later studies[100,101] established a linkage group for the porcine *HAL* gene, the designation of the MH gene giving rise to halothane sensitivity, *GPI,* and the gene for 6-phosphogluconate dehydrogenase (*PGD*), localized near the centromere of pig chromosome 6.[102-104] The homologous region around the human *GPI* locus was found on the long arm of chromosome 19,[105] making this a candidate region for human MH localization. Cloning of the human skeletal muscle ryanodine receptor (*RYR1*) cDNA[64] led to the localization of *RYR1* to human chromosome 19q13.1, in the same region as human *GPI*.[106]

Cloning of human *RYR1* permitted the identification of several restriction fragment length polymorphisms (RFLPs) in the human *RYR1* gene.[107] In a study of linkage between inheritance of MH, *RYR1* polymorphisms, and flanking markers, cosegregation with *RYR1* markers was found in 23 meioses in 9 families, leading to a lod score of 4.2 favoring linkage for a recombinant fraction of 0.0. The probability of linkage of more than 10,000 to 1 identified *RYR1* as a candidate gene for MH in humans.[107] In an independent study of linkage of human MH to a series of chromosome 19q markers, the MH locus was also assigned to the region of human chromosome 19 where *RYR1* was localized.[108]

3.4. MALIGNANT HYPERTHERMIA MUTATIONS IN SWINE

The demonstration of linkage of *RYR1* to MH led to parallel searches in both swine and humans for sequence differences in the *RYR1* gene between MH and

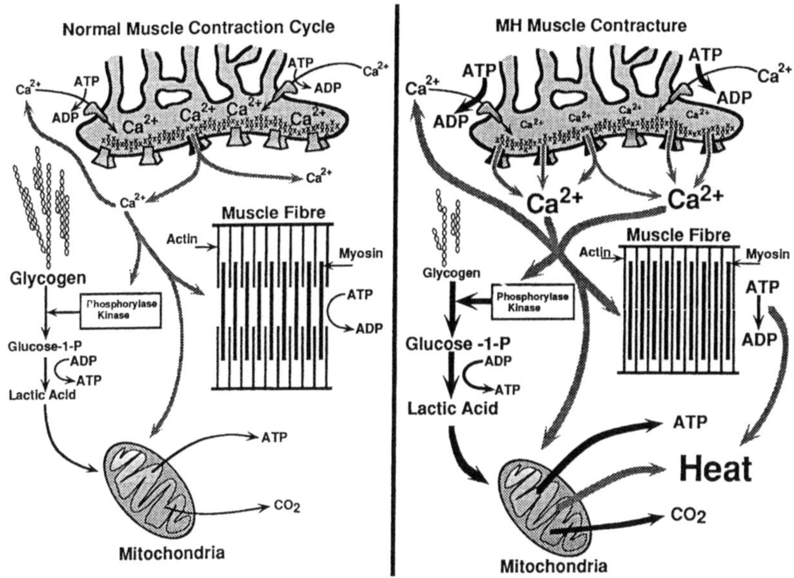

FIGURE 1. A proposed mechanism for induction of malignant hyperthermia due to abnormalities in the Ca^{2+} release channel of skeletal muscle sarcoplasmic reticulum. Muscle contraction, glycolysis, and mitochondrial function are regulated by cytoplasmic Ca^{2+} concentrations. In a normal relaxation-contraction cycle (left), Ca^{2+} is pumped into the sarcoplasmic reticulum by a Ca^{2+} ATPase to initiate relaxation, stored within the lumen in association with calsequestrin, and released through a Ca^{2+} release channel to initiate contraction. Glycolytic and aerobic metabolism proceed only rapidly enough to maintain the energy balance of the cell. The Ca^{2+} release channel can be regulated by physical interaction with the DHP receptor, or by Ca^{2+}, ATP, Mg^{2+}, and calmodulin. Even when stimulated, it has a relatively short open time. The abnormal malignant hyperthermia Ca^{2+} release channel (right) may be sensitive to lower concentrations of stimulators of opening, it releases Ca^{2+} at enhanced rates, and does not close readily. In the presence of anesthetic agents, the abnormal channel floods the cell with Ca^{2+} and overpowers the Ca^{2+} pump in its attempts to lower cytoplasmic Ca^{2+}. Chronically high levels of Ca^{2+} stimulate muscle contracture and glycolysis, accounting for rigidity and the generation of lactic acid. Damage to cell membranes and imbalances of ion transport resulting from lactic acid production and lowered ATP concentration can account for the life-threatening systemic problems which appear during a malignant hyperthermia episode. Heat is generated by the hydrolysis of ATP by the contracted muscle, and the Ca^{2+} pumps. Regeneration of ATP through aerobic metabolism accounts for CO_2 generation and enhanced O_2 uptake. (Adapted from MacLennan, D. H. and Phillips, M. S., *Science,* 256, 789, 1992. With permission.)

normal individuals.[65,109-112] In a comparison of the cDNA sequences of MH (Pietrain) and normal (Yorkshire) pigs, the substitution of T for C at nucleotide 1843, leading to the substitution of Cys for Arg at amino acid residue 615 (Table 1), was the only amino acid sequence alteration found.[65] This mutation resulted in the loss of a *Hin*PI restriction endonuclease site and the gain of a

TABLE I
***RYR1* Mutations Associated with MH or CCD**

Amino acid substitution	Nucleotide substitution	Detection	Association	Ref.
Cys for Arg^{163}	T for C487	Loss of *Bst* UI	MH, CCD	115
Arg for Gly^{248}	A for G742	Allele specific PCR	MH	111
Arg for Gly^{341}	A for G1021	SSCP	MH	116
Met for Ile^{403}	G for C1209	Loss of *Mbo*I	MH, CCD	115
Ser for Tyr^{522}	C for A1565	SSCP	MH,CCD	117
Cys for Arg^{614}	T for C1840	Loss of *Rsa*I	MH	110
Cys for Arg^{615} (Pig)	T for C1843	Loss of *Hin*PI, Gain of *Hgi*AI	MH, PSS	65
Arg for Gly^{2433}	A for G7297	Gain of *Dde*1	MH	73, 74
His for Arg^{2434}	A for G7301	Loss of *Hga*I	MH, CCD	72

*Hgi*AI site, making it possible to analyze the mutation by restriction endonuclease digestion or by allele-specific oligonucleotide hybridization analysis. Initial studies showed an association between inheritance of the mutation and inheritance of MH in some 80 animals from 5 different breeds. Tight linkage was established in a study of backcrosses between British Landrace heterozygous animals of the N/n genotype and homozygous MH animals of the *n/n* genotype[109] in which 376 animals were tested, including 338 representing informative meioses. Cosegregation of the phenotype with the Cys for Arg^{615} substitution was complete, leading to a lod score favoring linkage of 102 for a recombinant fraction of 0.0.

The fact that the identical mutation appeared in five lean, heavily muscled pig breeds led to the possibility that the mutation arose in a founder animal. Haplotype and genotype analysis using three markers covering over 100 kb within the *RYR1* gene showed the association of a specific haplotype with the MH phenotype. These results suggest that the disease originated in a founder animal and was selected for in breeding stock.

The disease gene seems to have been selected because it contributes to dressed carcass weight, to leanness and heavy muscling, and to the redistribution of fat from muscle to backfat.[35-37] A possible explanation for these phenomena is that a leaky or hypersensitive Ca^{2+} release channel could give rise to spontaneous muscle contraction, and the continual toning of the muscle may lead to muscle hypertrophy.[112] The utilization of ATP for spontaneous contraction would limit the deposition of fat.

3.5. MALIGNANT HYPERTHERMIA MUTATIONS IN HUMANS

The demonstration of linkage between MH and a substitution of Cys for Arg^{615} in the *RYR1* gene in swine led to a search for the corresponding mutation in human MH families. The equivalent mutation, Cys for Arg^{614}, was found in a single family of five members in which the mutation segregated with MH.[110] The mutation results in the loss of an *Rsa*1 restriction endonuclease site (Table 1), allowing its diagnosis by PCR amplification and cleavage.[45] This mutation has been found in about 2% of MH families worldwide. In most families it segregates with individuals who have been diagnosed by CHCT as MHS,[110,113] but in other families linkage is not observed. The question arises, therefore, whether the Arg^{614} to Cys mutation is really causal of MH, whether the phenotypic diagnosis is inaccurate, or whether other MH-causing mutations are also cosegregating in the family.

The assessment of the Arg^{614} to Cys mutation as causative of MH has a strong genetic and biochemical basis. The mutation has been linked to porcine MH with a lod score of 102 favoring linkage at θ max = 0.0. The corresponding mutation was found across a species barrier between swine and humans where it also segregates with MH in many families. In biochemical studies, a measurable defect was observed in closing of the Ca^{2+} release channel.[94,95] When expressed in muscle cells, the mutant form of the ryanodine receptor was found to release Ca^{2+} in response to lower levels of added caffeine or halothane than the normal channel.[114] Thus, when linkage of the Arg^{614} to Cys mutation to MH in humans cannot be demonstrated, it is likely that the cause is inaccurate diagnosis of MH through the CHCT. This problem of inaccuracy in the CHCT is a complicating factor in proving the causal nature of other MH mutations through linkage of inheritance of MH and mutant *RYR1* alleles.

Analysis of the DNA sequence of ryanodine receptors from probands from MH families has associated MH with mutations of Arg^{163} to Cys [115], of Gly^{248} to Arg [111], of Gly^{341} to Arg [116], of Ile^{403} to Met [115], of Tyr^{522} to Ser [117], of Gly^{2433} to Arg [73,74], and of Arg^{2434} to His [72] (Table 1). The mutation of Gly^{341} to Arg has been found to segregate with MH susceptibility in about 10% of European MH families, but in only 1% of Canadian MH families (unpublished).[116] In the seven families in which the mutation was found, six MHE individuals were noted. Of these, five did not contain the Gly^{341} to Arg mutation, suggesting that the majority of patients diagnosed as MHE by the European CHCT may not be genetically susceptible to MH. The mutation of Arg^{2433} to Gly is found in about 4% of MH families. The mutation of Arg^{163} to Cys has been found in less than 2% of MH families.[115] The remainder of the mutations have been found in only single families. Thus, the causal MH mutations in over 80% of MH families have yet to be found.

The search for additional MH mutations in *RYR1* is hampered largely by the size of the cDNA and of the gene from which it is derived. Sequencing of cDNA, or the combination of sequencing with other methods for detecting polymorphisms prior to sequencing, is feasible, but is laborious and expensive. Sequencing from genomic DNA will be fully feasible since 105 exon/intron boundaries have been sequenced. Data from phage cloning and from the isolation of YAC and cosmid clones containing the gene[118] show that the gene covers about 205 kbp. While the 15,000 bp cDNA encoding the 564,000 Da protein sequence is among the longest known, it is only 1/10 the size of the gene encoding dystrophin, mutant forms of which cause Duchenne muscular dystrophy.[119]

3.6. SEARCHING FOR A SECOND MHS LOCUS

On the basis of linkage to chromosome 19 and to mutations in *RYR1*, *RYR1* is established as a very strong candidate gene for MH, but it is not necessarily the only candidate gene for MH. There are cases where diagnosis appears to be accurate and where no linkage between *RYR1* and MH can be defined.[120-122] In at least some of these cases, both false positive and false negative diagnoses would have to be invoked to prove linkage. The finding of genetic heterogeneity is perhaps not surprising in light of reports that patients with other muscle diseases are subject to MH episodes or to positive CHCT.[6-15,123] Thus, individuals with central core disease, myotonia congenita, myotonia dystrophica, limb-girdle muscular dystrophy, Brody's disease, or Duchenne or Beckers muscular dystrophy have apparently had MH reactions or have been diagnosed as MHS by CHCT. Such a reaction could be postulated to result from a normal Ca^{2+} release channel which is triggered to open by a rise in Ca^{2+} in a muscle cell in which poor Ca^{2+} regulation exists due to defects in the Ca^{2+} pump, calsequestrin, or dihydropyridine receptor subunits, for example, or from increased membrane permeability due to secondary causes.[124] Thus, other defective proteins leading to poor Ca^{2+} regulation within the cell may eventually be shown to give rise to other forms of MH susceptibility. In some laboratories, as few as 35 to 50% of MH families[125] can be linked to chromosome 19q13.1, stimulating the search for a second MH gene locus.

Levitt et al.[126] and Olckers et al.[127] presented evidence for a second MHS locus on chromosome 17q21. There are three potential candidate genes on chromosome 17, the ß-subunit[128] and the γ-subunit[129] of the dihydropyridine receptor and a sodium channel gene, SCN4A, which has been linked to hyperkalemic periodic paralysis.[130] Other groups have attempted to confirm the localization of a second MHS locus on chromosome 17, but without success, and have ruled out the β- and γ-subunits of the DHP receptor as candidate genes for MH.[129,131] The α1-subunit of the DHP receptor has been localized to chromosome 1.[132] It has been excluded as a candidate gene for MH susceptibility,[132] but mutations in it have been linked to hypokalemic periodic paralysis in several families.[133] The α2-subunit of the DHP receptor has been localized to chromosome 7 [134] and a single MH family has been linked to the same region of chromosome 7, making the α2-subunit of the DHP receptor a candidate gene for MH. As yet, no mutation in this gene has been associated with MH.

In a collaborative study with Généthon, members of the European MH Group are attempting to link polymorphic markers covering the entire human genome[135] with MH. They have chosen 3 generation MH pedigrees which do not link to chromosome 19 and which are capable of generating a lod score of 3 or more. Using this methodology, linkage of a single MH family to chromosome 3q13.1 has been found.[159]

4. CENTRAL CORE DISEASE

Central core disease is a rare, nonprogressive myopathy characterized by hypotonia and proximal muscle weakness and presenting in infancy.[136] Additional variable clinical features include pes cavus, kyphoscoliosis, foot deformities, congenital hip dislocation, and joint contractures.[137-140] Although symptoms may be severe, up to 40% of patients demonstrating central cores may be clinically normal.[140] Diagnosis is made on the basis of the lack of oxidative enzyme activity in central regions of skeletal muscle cells,[141] observed upon histological examination of biopsies. Electron microscopic analysis shows disintegration of the contractile apparatus ranging from blurring and streaming of the Z lines to total loss of myofibrillar structure.[142,144,145] The sarcoplasmic reticulum and transverse tubular systems are greatly increased in content and, in general, less well structured.[144] Mitochondria are depleted in the cores, but may be enriched around the surfaces of the cores. Genetic analysis indicates that the disorder is inherited as an autosomal dominant trait with variable penetrance.[142,143,145,146]

An important feature of CCD is its close association with susceptibility to malignant hyperthermia.[6,13,146,147] This association led investigators to establish linkage between CCD and markers in the long arm of chromosome 19 in large Australian[148] and European[149] CCD pedigrees. A lod score of 11.8 favoring linkage with a recombinant fraction of 0.0, using markers within the *RYR1* gene, was also established in the Australian family.[150]

Analysis of *RYR1* cDNA sequences in several CCD families has led to the discovery of four mutations that are linked to CCD and/or MH (Table 1). Zhang et al.[72] linked the substitution of Arg^{2434} with His to CCD in a Canadian family, obtaining a lod score of 4.8 favoring linkage with a recombinant fraction of 0.0. This mutation was found in only a single pedigree of more than a dozen examined. Quane et al.[115] found two mutations in small CCD families that were linked to inheritance of either MH or CCD in these families. These mutations were the substitution of Met for Ile^{403} and the substitution of Cys for Arg^{163}. The Arg^{163} to Cys mutation gave rise to invariant MH susceptibility, but to variable formation of central cores. The Ile^{403} to Met mutation was found in a single CCD pedigree.

It is of interest that 6 out of 8 MH or CCD mutations lie between amino acids 163 and 614 (Table 1), in a region homologous to the IP_3 binding region in the IP_3 receptor.[151,152] Such a cluster of mutations would suggest that this region of the molecule forms a regulatory domain in the Ca^{2+} release channel. A second MH regulatory domain is found around amino acid residues 2433/2434.[72-74] Studies of tryptic digestion[153] suggest that proteolytic cleavage sites occur at about residues 500 and 1400, the first site lying in the midst of the MH mutation cluster. These observations may indicate that two interacting regulatory domains lie within the first 1400 residues of the Ca^{2+} release channel. A second interesting feature of the known MH mutations is that six of the eight involve either loss or gain of an Arg residue. This suggests that positive changes within the two MH domains are critical to regulatory function.

The discovery of mutations in *RYR1* potentially causal of CCD and MH raises interesting questions concerning the pathophysiology of these diseases. It is clear that, in both cases, Ca^{2+} regulation is imbalanced, leading to the contracture and hypermetabolism that characterize MH. In some cases, however, the altered Ca^{2+} regulation leads to disorganization of the contractile proteins in the central core, a proliferation of sarcoplasmic reticulum and transverse tubules, and a loss of functional mitochondria. It is possible that these structural alterations arise developmentally and are the result of disorganization imposed on the developing fiber by alterations in the physical properties of the Ca^{2+} release channel, which may play a critical role in the organization of the sarcotubular membrane system.

Another possibility is that these alterations occur subsequent to the initial formation of a fully functional muscle fiber and are the result of physiological adaptation to functional alterations in the channel that lead to elevated Ca^{2+} levels within the myofibril.[72] Myofibrils regulate Ca^{2+} through at least four systems, Ca^{2+} pumps and Na/Ca^{2+} exchangers in the plasma membrane can remove Ca^{2+} from the muscle cell.[154] Of these, the Ca^{2+} pump has the higher affinity for Ca^{2+}. The sarcoplasmic reticulum is the major regulator of Ca^{2+} within the muscle cell, removing it from the cytoplasm, storing it, and releasing it again to initiate muscle contraction. If Ca^{2+} concentrations are elevated, the mitochondria can transport Ca^{2+} to matrix spaces, thereby protecting the cell from Ca^{2+}-induced damage.[155] If the Ca^{2+} release channel were to release excessive amounts of Ca^{2+} within the muscle cell, then the Na^+/Ca^{2+} exchanger and mitochondria might play a more important role in Ca^{2+} regulation in a CCD cell than in a normal cell. Extrusion of excess Ca^{2+} from the cell might, itself, have deleterious effects on the skeletal muscle cell, which is believed to carry out intracellular cycling of a constant level of Ca^{2+} rather than the fluxing of Ca^{2+} from external sources.

Pumps and exchangers in the plasma membrane might be more effective in protecting the periphery of the cell than the interior of the cell, where the full burden of regulation of excess Ca^{2+} would fall on the sarcoplasmic reticulum and mitochondria. It is possible that mitochondria, which have a high capacity for Ca^{2+} uptake and would, undoubtedly, participate in removal of excess Ca^{2+} from central areas of the cell, might destroy themselves in an effort to protect the cell from Ca^{2+}-induced necrosis.[155] Loss of mitochondria from the center of the cell would, in turn, lead to lower ATP synthesis and might be an underlying cause of the disorganization of the central core, leading to muscle weakness and muscle atrophy. Elevated Ca^{2+} in the interior of the muscle cell might have the same effects on the core of the muscle cell as MH would have on the whole cell. Of most interest would be its effects on muscle contraction. The differential contraction of the core of the muscle, in relation to the periphery, could lead to the disorganization and 'streaming' of both fibers and membrane systems that is observed in the central cores. The profusion of sarcoplasmic reticulum and transverse tubules might be induced at the gene level by high local Ca^{2+} concentrations or their downstream effects (Figure 2).

It is of interest that the Arg^{615} to Cys mutation is associated with muscle hypertrophy in swine while the Arg^{2434} to His, Tyr^{522} to Ser, Ile^{403} to Met, and Arg^{163} to Cys mutations are associated with variable degrees of muscle atrophy, metabolically inert cores, and proximal muscle weakness. If all of these mutations led to poorly regulated Ca^{2+} release into the muscle cell, they could trigger spontaneous muscle contractions. Such spontaneous contractions could lead to the muscle hypertrophy observed in swine. In this case, the system of pumps and exchangers in the plasma membrane and organelle systems of mitochondria and sarcoplasmic reticulum within the cell could remove excess Ca^{2+} from the sarcoplasm without deleterious effects on the muscle cell. As we have outlined above, however, the CCD mutations might be more severe, leading to damage to the interior of the cell and to loss of mitochondrial function and structural abnormalities in the central core. These, in turn, could lead to muscle weakness and atrophy. Thus, mutations in *RYR1* can lead to a spectrum of pathophysiological responses ranging from muscle hypertrophy to muscle atrophy.

5. KING-DENBOROUGH SYNDROME

Anesthetic-induced malignant hyperthermia was reported by King and Denborough[7] to occur in children with particular congenital abnormalities such as short stature, scoliosis, pectus deformity, delay in motor development, ptosis, low-set ears, anti-Mongolian slanted eyes, and cryptorchidism.[156] This syndrome is inherited as an autosomal dominant trait. It is not yet known whether the syndrome is linked to chromosome 19 and whether it could result from defects in *RYR1*.

6. CONCLUSIONS

Research over the past quarter-century has led to a very good understanding of the physiological and genetic basis for malignant hyperthermia in humans and in swine, but many goals are yet to be attained. Perhaps the most important immediate goal is to define all of the genes and all of the mutations in those genes which are causal of human MH. Once most of the MH mutations are known, it should be possible to follow inheritance of these mutations and to designate those members of MH families who are MHS. This will assure that MHS individuals will receive a regimen of "safe" anesthetics, while normal family members can utilize more conventional anesthetics. A second important goal is to utilize the MH gene in ways that will be of most benefit to the pork industry. It is feasible to identify and remove the MH gene from the porcine population within a very short time, if that should prove desirable.[157] On the

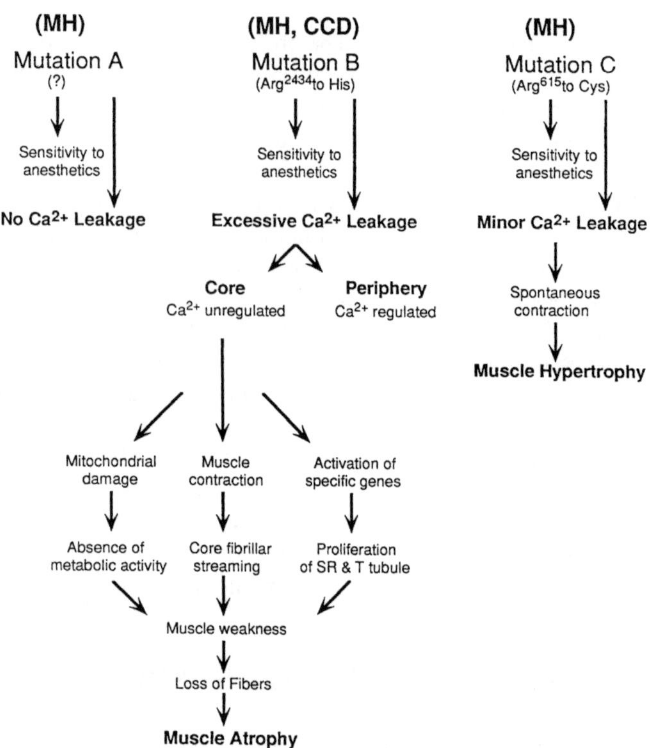

FIGURE 2. A proposed mechanism for the differential phenotypic effects of different MH and CCD mutations. MH mutations in the ryanodine receptor lead to the common phenotype of sensitivity to anesthetics. Some mutations may also lead to spontaneous Ca^{2+} release sufficient to trigger spontaneous contractions. If this trigger Ca^{2+} were readily regulated, the major phenotypic effect would be spontaneous exercise-induced muscle hypertrophy. The Arg^{615} to Cys mutation was selected in swine because it leads to increased lean muscle mass. Mutations leading to excessive spontaneous Ca^{2+} release may have no phenotypic effect on the periphery of the cell, but be deleterious to the central core. Ca^{2+} released from the sarcoplasmic reticulum can be regulated by four systems including the organellar sarcoplasmic reticulum and mitochondria and the plasma membrane Ca^{2+} pumps and Na^+/Ca^{2+} exchangers. Under normal circumstances, the bulk of the Ca^{2+} is cycled only through the sarcoplasmic reticulum. If enhanced Ca^{2+} release occurred spontaneously, in CCD muscle, the additional Ca^{2+} regulatory systems might be co-opted to regulate Ca^{2+}. The plasma membrane exchangers and pumps would be effective in regulating Ca^{2+} near the periphery but, in the core, mitochondria may be forced to bear this load, destroying themselves in the process. The degeneration of mitochondria could lead to the degeneration of a central, possibly compartmented, core. The disorganization of the central core might be brought about by higher core levels of Ca^{2+}, which could cause contraction at the core and lead to myofibrillar streaming and membrane disorganization. Elevated Ca^{2+} may stimulate proliferation of internal membrane proteins at the transcriptional level. The phenotypic effects would be the formation of a disorganized, metabolically deficient core which could lead to cell death and muscle atrophy.

other hand, since the gene is beneficial in terms of food conversion, leanness, and muscularity, research is underway to evaluate the usefulness of the gene to the pork industry. A third goal will be to understand the functional consequences of the structural alterations in the Ca^{2+} release channel that lead to its involvement in MH episodes.[158] Very little is known about the structural and functional domains of the protein and how they affect gating of the channel. MH mutations are clustered between residues 163 and 614 in a domain that corresponds to the IP3 binding domain of the IP_3 receptor. They are also appearing in a predicted regulatory domain which begins at about residue 2433. These mutations frequently involve the basic residue, arginine. Thus studies of the regions in which MH mutations are clustered are leading to the recognition of important regulatory domains in the Ca^{2+} release channel. A fourth goal will be to understand, at the physiological level, how MH relates to additional abnormalities in muscle cells, such as the formation of central cores, and how other myopathies can underlie anesthetic-induced malignant hyperthermia.[13] The question of how central cores form in muscle from a mutation that also causes MH raises the important question of how chronic variations in Ca^{2+} levels in a muscle cell can affect its physiology.

In this chapter, we have described our current level of understanding of the cause of an abnormality that is important both to human medicine and to the agricultural industry. Understanding of the genetic basis for the abnormality will ultimately provide not only accurate diagnosis for human carriers of the defective gene but will also be of economic benefit to the pork industry.

ACKNOWLEDGMENTS

We thank our many colleagues for advice and discussion in the preparation of this review. Research grants to D.H.M., supporting original work from our laboratory, were from the Medical Research Council of Canada (MRCC), the Muscular Dystrophy Association of Canada (MDAC), the Heart and Stroke Foundation of Ontario (HSFO), and the Canadian Genetic Diseases Network of Centers of Excellence. M.S.P. is a predoctoral fellow of the HSFO and of the MRCC.

REFERENCES

1. **Britt, B. A.,** Malignant hyperthermia: a review, in *Thermoregulation: Pathology, Pharmacology and Therapy*, Schonbaum, E. and Lomax, P., Eds., Pergamon Press, New York, 1991, 179.
2. **Denborough, M. A. and Lovell, R. R. H.,** Anesthetic deaths in a family, *Lancet,* ii, 45, 1960.
3. **Denborough, M. A., Forster, J. F. A., and Lovell, R. R. H.,** Anesthetic deaths in a family, *Br. J. Anaesth.,* 34, 395, 1962.

4. **Britt, B. A. and Kalow, W.,** Malignant hyperthermia: a statistical review, *Can. Anaesth. Soc. J.,* 17, 293, 1970.
5. **Ording, H.,** Incidence of malignant hyperthermia in Denmark, *Anesth. Analg.*, 64, 700, 1985.
6. **Denborough, M. A., Dennett, X., and Anderson, R. McD.,** Central-core disease and malignant hyperpyrexia, *Br. Med. J., 1*, 272, 1973.
7. **King, J. O. and Denborough, M. A.,** Anaesthetic-induced malignant hyperpyrexia in children, *J. Pediatr.,* 83, 37 1973.
8. **Isaacs, H. and Badenhorst, M. E.,** Dominantly inherited malignant hyperthermia (MH) in the King-Denborough syndrome, *Muscle Nerve, 15*, 740, 1992.
9. **Brownell, A. K. W., Paasuke, R. T., Elash, A., Fowlow, S. B., Seagram, C. G. F., Diewold, R. J., and Friesen, C.,** Malignant hyperthermia in Duchenne muscular dystrophy, *Anesthesiology*, 58, 180, 1983.
10. **Kefler, H. M., Singer, W. D., and Reynolds, R. N.,** Malignant hyperthermia in a child with Duchenne muscular dystrophy, *Pediatrics*, 71, 1, 118, 1983.
11. **Sethna, N. F. and Rockoff, M. A.,** Cardiac arrest following inhalation induction of anesthesia in a child with Duchenne's muscular dystrophy, *Can. Anaesth. Soc. J.*, 33, 799, 1986.
12. **Delphin, E., Jackson, D., and Rothstein, P.,** Use of succinylcholine during elective pediatric anesthesia should be reevaluated, *Anesth. Analg.*, 66, 1190, 1987.
13. **Brownell, A. K. W.,** Malignant hyperthermia: relationship to other diseases, *Br. J. Anaesth.,* 60, 303, 1988.
14. **Heiman-Patterson, T., Rosenberg, H., Fletcher, J. E., and Tahmoush, A. J.,** Malignant hyperthermia in myotonia congenita, Halothane-caffeine contracture testing in neuromuscular disease, *Muscle Nerve,* 11, 453, 1988.
15. **Karpati, G., Charuk, J., Carpenter, S., Jablecki, C., and Holland, P.,** Myopathy caused by a deficiency of Ca^{2+}-adenosine triphosphatase in sarcoplasmic reticulum (Brody's disease), *Ann. Neurol.,* 20, 38, 1986.
16. **Kalow, W., Britt, B. A., and Terreau, M. E.,** Metabolic error of muscle metabolism after recovery from malignant hyperthermia, *Lancet,* ii, 895, 1970.
17. **Ellis, F. R. and Harriman, D. G. F.,** A new screening test for susceptibility to malignant hyperpyrexia, *Br. J. Anaesth.,* 45, 638, 1973.
18. **Larach, M. G.,** The North American Malignant Hyperthermia Group, Standardization of the caffeine halothane muscle contracture test, *Anesth. Analg.,* 69, 511, 1989.
19. **European MH Group,** Malignant hyperpyrexia, a protocol for the investigation of malignant hyperthermia (MH) susceptibility, *Br. J. Anaesth.,* 56, 1267, 1984.
20. **Larach, M. G.,** Should we use muscle biopsy to diagnose malignant hyperthermia susceptibility, *Anesthesiology*, 79, 1, 1993.
21. **Larach, M. G., Landis, J. R., Shirk, B. S., and Diaz, M.,** Prediction of malignant hyperthermia susceptibility in man, improving sensitivity of the caffeine halothane contracture test, *Anesthesiology*, 77, A1052, 1992.
22. **Larach, M. G., Landis, J. R., Bunn, J. S., and Diaz, M.,** Prediction of malignant hyperthermia susceptibility in low-risk subjects; an epidemiologic investigation of caffeine halothane contracture responses, *Anesthesiology,* 76, 16, 1992.
23. **Isaacs, H. and Badenhorst, M.,** False-negative results with muscle caffeine halothane contracture testing for malignant hyperthermia, *Anesthesiology*, 79, 5, 1993.
24. **MacKenzie, A. E., Allen, G., Lahey, D., Crossan, M. L., Nolan, K., Mettler, G., Worton, R.G., MacLennan, D. H., and Korneluk, R.,** A comparison of the caffeine halothane muscle contracture test with the molecular genetic diagnosis of malignant hyperthermia, *Anesthesiology*, 75, 4, 1991.
25. **Larach, M. G., Localio, A. R., Allen, G. C., Denborough, M. A., Ellis, F. R., Gronert, G. A., Kaplan, R. F., Muldoon, S. M., Nelson, T. E., Ording, H., Rosenberg, H., Waud, B. E., and Wedel, D. J.,** A clinical grading scale to predict malignant hyperthermia susceptibility, *Anesthesiology*, 80, 771, 1994.

26. **Hopkins, P. M., Ellis, F. R., and Halsall, P. J.,** Ryanodine contracture: a potentially specific *in vitro* diagnostic test for malignant hyperthermia, *Br. J. Anaes.*, 66, 611, 1991.
27. **Lenzen, C., Roewer, N., Wappler, F., Scholz, J., Kahl, J., Blank, M., Rumberger, E., and Schulte, am esch, J.,** Accelerated contractures after administration of ryanodine to skeletal muscle of malignant hyperthermia susceptible patients, *Br. J. Anaes.*, 71, 242, 1993.
28. **Klip, A., Britt, B. A., Elliott, M. E., Pegg, W., Frodis, W., and Scott, E.,** Anesthetic-induced increase in ionised calcium in blood mononuclear cells from malignant hyperthermia patients, *Lancet*, 463, 1987.
29. **Klip, A., Ramlal, T., Walker, D., Britt, B. A., and Elliott, M. E.,** Selective increase in cytoplasmic calcium by anesthetic in lymphocytes from malignant hyperthermia-susceptible pigs, *Anesth. Analg.*, 66, 381, 1987.
30. **Lopez-Padrino, J. R.,** Free calcium concentration in skeletal muscle of malignant hyperthermia susceptible subjects, effects of ryanodine, in *Malignant Hyperthermia: A Genetic Membrane Disease*, Ohnishi, S. T. and Ohnishi, T., Eds., CRC Press, Boca Raton, FL, 1993, chap. 11.
31. **Iaizzo, P., Klein, W., and Lehman-Horn, F.,** Fura 2 detected myoplasmic calcium and its correlation with contracture force in skeletal muscle from normal and malignant hyperthermia susceptible pigs, *Pflugers Arch.*, 411, 648, 1988.
32. **Payen, J., Bosson, J., Bourdon, L., Jacquout, C., Le Bas, J., Steiglitz, P., and Benahid, A.,** Improved noninvasive diagnostic testing for malignant hyperthermia susceptibility from a combination of metabolites determined *in vivo* with ^{31}P-magnetic resonance spectroscopy, *Anesthesiology*, 78, 848, 1993.
33. **Hall, L. W., Woolf, N., Bradley, J. W., and Jolly, D. W.,** Unusual reaction to suxamethonium chloride, *Br. Mol. J.*, 2, 1305, 1966.
34. **Harrison, G. G.,** Porcine malignant hyperthermia, *Int. Anesthesiol. Clin.*, 17, 25, 1979.
35. **Simpson, S. P. and Webb, A. J.,** Growth and carcass performance of British Landrace pigs heterozygous at the halothane locus, *Anim. Prod.*, 49, 503, 1989.
36. **Webb, A. J. and Simpson S. P.,** Performance of British Landrace pigs selected for high and low incidence of halothane sensitivity 2 Growth and carcass traits, *Anim. Prod.*, 43, 493, 1986.
37. **O'Brien, P. J., Ball, R. O., and MacLennan, D. H.,** Effects of heterozygosity for the mutation causing porcine stress syndrome on carcass quality and live performance characteristics, in *Proc. 13th Int. Pig Vet. Soc. Congr.*, Bangkok, 1994, 481,
38. **O'Brien, P. J.,** Porcine malignant hyperthermia susceptibility. Hypersensitive calcium-release mechanism of skeletal muscle sarcoplasmic reticulum, *Can. J. Vet. Res.*, 50, 318, 1986b.
39. **Webb, A. J., Imlah, P., and Carden, A. E.,** Succinylcholine and halothane as a field test for the heterozygote at the halothane locus in pigs, *Anim. Prod.*, 42, 275, 1986.
40. **Seeler, D. C., McDonell, W. N., and Basrur P. K.,** Halothane and halothane/succinylcholine induced malignant hyperthermia (porcine stress syndrome) in a population of Ontario boars, *Can. J. Comp. Med.*, 47, 284, 1983.
41. **Gallant, E. M., Mickelson, J. R., Roggow, B. D., Donaldson, S. K., Louis, C. F., and Rempel, W. E.,** Halothane-sensitivity gene and muscle contractile properties in malignant hyperthermia, *Am. Physiol. Soc.*, C781, 1989.
42. **Nelson, T. E., Flewellen, E. H., and Gloyna, D. F.,** Spectrum of susceptibility to malignant hyperthermia — diagnostic dilemma, *Int. Anest. Res. Soc.*, 62, 545, 1983.
43. **Fletcher, J. E., Calvo, P. A., and Rosenberg, H.,** Phenotypes associated with malignant hyperthermia susceptibility in swine genotyped as homozygous or heterozygous for the ryanodine receptor mutation, *Br. J. Anaes.*, 71, 410, 1993.
44. **Eikelenboom, G. and Minkema, D.,** Prediction of pale, soft, exudative muscle with a non-lethal test for halothane induced porcine malignant hyperthermia syndrome, *Neth. J. Vet. Sci.*, *99*, 421, 1974.

45. **Otsu, K., Phillips, M. S., Khanna, V. K., de Leon, S., and MacLennan, D. H.,** Refinement of diagnostic assays for a probable causal mutation for porcine and human malignant hyperthermia, *Genomics,* 13, 835, 1992.
46 **Pommier, S. A., Houde, A., Rousseau, F., and Savoie, Y.,** The effect of malignant hyperthermia as determined by a restriction endonuclease assay on carcass characteristics of commercial crossbred pigs, *Can. J. Anim. Sci.,* 72, 973, 1992.
47. **Pommier, S. A. and Houde, A.,** Effect of the genotype for malignant hyperthermia as determined by a restriction endonuclease assay on the quality characteristics of commercial pork loins, *J. Anim. Sci.*, 71, 420, 1993.
48. **Fortin, A., Pommier, S. A., and Houde, A.,** PSE in pork: the relationship between genotype as determined by the restriction endonuclease assay, and the environment, *ICoMST*, 2, 173, 1993.
49. **Ebashi, S.,** Third component participating in the superprecipitation of "natural actomyosin", *Nature*, 200, 1010, 1963.
50. **Zot, A. S. and Potter, J. D.,** Structural aspects of troponin-tropomyosin regulation of skeletal muscle contraction, *Annu. Rev. Biophys. Biophys. Chem.*, 16, 535, 1987.
51. **Brostrom, C. O., Hunkeler, F. L., and Krebs, E. G.,** The regulation of skeletal muscle phosphorylase kinase by Ca^{2+}, *J. Biol. Chem.*, 246, 1961, 1971.
52. **Berman, M. C., Harrison, G. G., Bull, A. B., and Kench, J. E.,** Changes underlying halothane-induced malignant hyperthermia in Landrace pigs, *Nature*, 225, 653, 1970.
53. **Endo, M.,** Calcium release from the sarcoplasmic reticulum, *Physiol. Rev.*, 57, 71, 1977.
54. **Weber, A. and Herz, R.**, The relationship between caffeine contracture of intact muscle and the effect of caffeine on reticulum, *J. Gen. Physiol.,* 52, 750, 1968.
55. **Inui, M., Saito, A., and Fleischer, S.,** Purification of the ryanodine receptor and identity with feet structures of junctional terminal cisternae of sarcoplasmic reticulum from fast skeletal muscle, *J. Biol. Chem.*, 262, 1740, 1987.
56. **Meissner, G.,** Adenine nucleotide stimulation of Ca^{2+} induced Ca^{2+} release in sarcoplasmic reticulum, *J. Biol. Chem.*, 259, 2365, 1984.
57. **Meissner, G.,** Evidence of a role for calmodulin in the regulation of calcium release from skeletal muscle sarcoplasmic reticulum, *Biochemistry*, 25, 424, 1986.
58. **Meissner, G.,** Ryanodine activation and inhibition of the Ca^{2+} release channel of sarcoplasmic reticulum, *J. Biol. Chem.*, 261, 6300, 1986.
59. **Meissner, G., Darling, E., and Eveleth, J.,** Kinetics of rapid Ca^{2+} release by sarcoplasmic reticulum. Effects of Ca^{2+}, Mg^{2+}, and adenine nucleotides, *Biochemistry*, 25, 236, 1986.
60. **Smith, J. S., Coronado, R., and Meissner, G.,** Sarcoplasmic reticulum contains adenine nucleotide activated calcium channels, *Nature*, 316, 446, 1985.
61. **Miyamoto, H. and Racker, E.,** Mechanism of calcium release from skeletal sarcoplasmic reticulum, *J. Membr. Biol.*, 66, 193, 1982.
62. **Morii, H. and Tonomura, Y.,** The gating behavior of a channel for Ca^{2+}-induced Ca^{2+} release in fragmented sarcoplasmic reticulum, *J. Biochem.*, 93, 1271, 1983.
63. **Takeshima, H., Nishimura, S., Matsumoto, T., Ishida, H., Kangawa, K., Minamino, N., Matsuo, H., Ueda, M., Hanaoka, M., Hirose, T., and Numa, S.,** Primary structure and expression from complementary DNA of skeletal muscle ryanodine receptor, *Nature*, 339, 439, 1989.
64. **Zorzato, F., Fujii, J., Otsu, K., Phillips, M., Green, N. M., Lai, F. A., Meissner, G., and MacLennan, D. H.,** Molecular cloning of cDNA encoding human and rabbit forms of the Ca^{2+} release channel (ryanodine receptor) of skeletal muscle sarcoplasmic reticulum, *J. Biol. Chem.*, 265, 2244, 1990.
65. **Fujii, J., Otsu, K., Zorzato, F., de Leon, S., Khanna, V. K., Weiler, J., O'Brien, P. J., and MacLennan, D. H.,** Identification of a mutation in porcine ryanodine receptor associated with malignant hyperthermia, *Science*, 253, 448, 1991.
66. **Otsu, K., Willard, H. F., Khanna, V. K., Zorzato, F., Green, N. M., and MacLennan, D. H.,** Molecular cloning of cDNA encoding the Ca^{2+} release channel (ryanodine receptor) of rabbit cardiac muscle sarcoplasmic reticulum, *J. Biol. Chem.*, 265, 13472, 1990.

67. **Nakai, J., Imagawa, T., Hakamata, Y., Shigekawa, M., Takeshima, H., and Numa, S.,** Primary structure and functional expression from cDNA of the cardiac ryanodine receptor/calcium release channel, *FEB Lett.*, 271, 169, 1990.
68. **Giannini, G., Clementi, E., Ceci, R., Marziali, G., and Sorrentino, V.,** Expression of a ryanodine receptor-Ca^{2+} channel that is regulated by TGF-ß, *Science*, 257, 91, 1992.
69. **Hakamata, Y., Nakai, J., Takeshima, H., and Imoto, K.,** Primary structure and distribution of a novel ryanodine receptor/calcium release channel from rabbit brain, *Fed. Europ. Biochem. Soc.*, 312, 229, 1992.
70. **Suko, J., Maurer-Fogy, I., Plank, B., Bertel, O., Wyskovsky, W., Hohenegger, M., and Hellmann, G.,** Phosphorylation of serine 2843 in ryanodine receptor-calcium release channel of skeletal muscle by cAMP-, cGMP- and CaM-dependent protein kinase, *Biochim. Biophys. Acta*, 1175, 193, 1993.
71. **Witcher, D. R., Kovacs, R. J., Schulman, H., Cefali, D. C., and Jones, L. R.,** Unique phosphorylation site on the cardiac ryanodine receptor regulates calcium channel activity, *J. Biol. Chem.*, 266, 11144, 1991.
72. **Zhang, Y., Chen, H. S., Khanna, V. K., de Leon, S., Phillips, M. S., Schappert, K., Britt, B. A., Brownell, A. K. W., and MacLennan, D. H.,** Identification of a mutation in human ryanodine receptor associated with central core disease, *Nature Genetics*, 5, 61, 1993.
73. **Phillips, M. S., Khanna, V. K., de Leon, S., Frodis, W., Britt, B. A., and MacLennan, D. H.,** The substitution of Arg for Gly2433 in the human skeletal muscle ryanodine receptor is associated with malignant hyperthermia, *Hum. Mol. Genet.*, 3, 2181, 1994.
74. **Keating, K. E., Quane, K. A., Manning, B. M., Lehane, M., Hartung, E., Censier, K., Urveyler, A., Klausnetger, M., Müller, C. R., Heffron, J. J. A., and McCarthy, T. V.,** Detection of a novel RYR1 mutation in four malignant hyperthermia pedigrees, *Hum. Mol. Genet.*, 3, 1855, 1994.
75. **McKinley, D. and Meissner, G.,** Evidence for a K^+, Na^+ permeable channel in sarcoplasmic reticulum, *J. Membr. Biol.*, 44, 159, 1978.
76. **Schneider, M. F. and Chandler, W. K.,** Voltage dependent charge movement in skeletal muscle, a possible step in excitation-contraction coupling, *Nature*, 242, 747, 1973.
77. **Rios, E. and Pizzaro, G.,** Voltage sensor of excitation-contraction coupling in skeletal muscle, *Physiol. Rev.*, 71, 849, 1991.
78. **MacLennan, D. H. and Wong P. T. S.,** Isolation of a calcium-sequestering protein from sarcoplasmic reticulum, *Proc. Natl. Acad. Sci. U.S.A.*, 68, 1231, 1971.
79. **Meissner, G.,** Isolation and characterization of two types of sarcoplasmic reticulum vesicles, *Biochem. Biophys. Acta*, 389, 51, 1975.
80. **Kim, K. C., Caswell, A. H., Talvenheimo, J. A., and Brandt, N. R.,** Isolation of a terminal cisternae protein which may link the dihydropyridine receptor to the junctional foot protein in skeletal muscle, *Biochemistry*, 29, 9281, 1990.
81. **Knudson, C. M., Stang, K. J., and Jorgensen, A. O.,** Biochemical characterization and ultrastructural localization of a major junctional sarcoplasmic reticulum glycoprotein (Triadin), *J. Biol. Chem.*, 268, 12637, 1993.
82. **Knudson, C. M., Stang, K. K., Moomaw, C. R., and Slaughter, C. A.,** Primary structure and topological analysis of a skeletal muscle-specific junctional sarcoplasmic reticulum glycoprotein (Triadin), *J. Biol. Chem.*, 268, 12646, 1993.
83. **Franzini-Armstrong, C.,** Studies of the triad. I. Structure of the junction in frog twitch fibers, *J. Cell. Biol.*, 47, 488, 1970.
84. **Block, B. A., Imagawa, T., Campbell, K. P., and Franzini-Armstrong, C.,** Structural evidence for direct interaction between the molecular junction in skeletal muscle, *J. Cell. Biol.*, 107, 2587, 1988.
85. **Tanabe, T., Beam, K. G., Adams, B. A., Niidome, T., and Numa, S.,** Regions of the skeletal muscle dihydropyridine receptor critical for excitation-contraction coupling, *Nature*, 346, 567, 1990b.

86. **Lu, X. and Meissner, G.,** Activation of the skeletal muscle calcium channel by a cytoplasmic loop of the dihydropyridine receptor, *J. Biol. Chem.*, 269, 6511, 1994.
87. **O'Brien, P. J.,** Porcine malignant hyperthermia susceptibility. Increased calcium sequestering activity of skeletal muscle sarcoplasmic reticulum, *Can. J. Vet. Res.*, 50, 329, 1986a.
88. **Nelson, T. E.,** SR function in malignant hyperthermia, *Cell*, 9, 257, 1988.
89. **O'Brien, P, J.,** Etiopathogenetic defect of malignant hyperthermia. Hypersensitive calcium-release channel of skeletal muscle sarcoplasmic reticulum, *Vet. Res. Commun.*, 11, 527, 1987.
90. **Endo, M., Yagi, S., Ishizuka, T., Horiuti, K., Koga, Y., and Amaha, K.,** Changes in the Ca-induced Ca release mechanism in sarcoplasmic reticulum from a patient with malignant hyperthermia, *Biomed. Res.*, 4, 83, 1983.
91. **Ohnishi, S. T., Taylor, S., and Gronert, G. A.,** Calcium-induced Ca^{2+} release from sarcoplasmic reticulum of pigs susceptible to malignant hyperthermia. The effects of halothane and dantrolene, *FEBS Lett.*, 161, 103, 1983.
92. **Nelson, T. E.,** Abnormality in calcium release from skeletal sarcoplasmic reticulum of pigs susceptible to malignant hyperthermia, *J. Clin. Invest.*, 72, 862, 1983.
93. **Kim, D. H., Sreter, F. A., Ohnishi, S. T., Ryan, J. F., Roberts, J., Allen, P. D., Meszaros, L. G., Antoniu, B., and Ikemoto, N.,** Kinetic studies of Ca^{2+} release from sarcoplasmic reticulum of normal and malignant hyperthermia susceptible pig muscles, *Biochim. Biophys. Acta*, 775, 320, 1984.
94. **Fill, M., Coronado, R., Mickelson, J. R., Vilven, J., Ma, J., Jacobson, B. A., and Louis, C. F.,** Abnormal ryanodine receptor channels in malignant hyperthermia, *Biophys. J.*, 50, 471, 1990.
95. **Shomer, N. H., Louis, C. F., Fill, M., Litterer, L. A., and Mickelson, J. R.,** Reconstitution of abnormalities in the malignant hyperthermia-susceptible pig ryanodine receptor, *Am. Physiol. Soc.*, C125, 1993.
96. **Fill, M., Stefani, E., and Nelson, T. E.,** Abnormal human sarcoplasmic reticulum Ca^{2+} release channels in malignant hyperthermia skeletal muscle, *Biophys. J.*, 59, 1085, 1991.
97. **Mickelson, J. R., Gallant, E. M., Litterer, L. A., Johnson, K. M., Rempel, W. E., and Louis, C. F.,** Abnormal sarcoplasmic reticulum ryanodine receptor in malignant hyperthermia, *J. Biol. Chem.*, 263, 9310, 1988.
98. **Knudson, C. M., Mickelson, J. R., Louis, C. F., and Campbell, K. P.,** Distinct immunopeptide maps of the sarcoplasmic reticulum Ca^{2+} release channel in malignant hyperthermia, *J. Biol. Chem.*, 265, 2421, 1990.
99. **Andersen, E. and Jensen, P.,** Close linkage established between the HAL locus for halothane sensitivity and the PHI (phosphohexose isomerase) locus in pigs of the Danish Landrace breed, *Nord. Vet. Med.*, 29, 502, 1977.
100. **Gahne, B. and Juneja, R. K.,** Prediction of the halothane (Hal) genotypes of pigs by deducing Hal, Phi, Po2, Pgd haplotypes of parents and offspring. Results from a large-scale practice in Swedish breed, *Anim. Blood Groups Biochem. Genet.*, 16, 265, 1985.
101. **Archibald, A. L. and Imlah, P.,** The halothane sensitivity locus and its linkage relationships, *Anim. Blood Groups Biochem. Genet.*, 16, 253, 1985.
102. **Davies, W., Harbitz, I., Fries, R., Stranzinger, G., and Hauge, J. G.,** Porcine malignant hyperthermia carrier detection and chromosomal assignment using a linked probe, *Anim. Genet.*, 19, 203, 1988.
103. **Chowdhary, B. P., Harbitz, I., Makinen, A., Davies, W., and Gustavvson, I.,** Localization of the glucose phosphate isomerase gene to p12-q21 segment of chromosome 6 in pig by *in situ* hybridization, *Hereditas*, 111, 73, 1989.
104. **Harbitz, L., Chowdhary, B., Thomsen, P., Davies, W., Kaufman, U., Kran, S., Gustavvson, I., Christensen, K., and Hauge, J.,** Assignment of the porcine calcium release channel gene, a candidate for the malignant hyperthermia locus, to the 6p11-q21 segment of chromosome 6, *Genomics*, 9, 243, 1990.

105. **Lusis, A. J., Heinzmann, C., Sparkes, R. S., Scott, J., Knott, T. J., Geller, R., Sparkes, M. C., and Mohandas, T.,** Regional mapping of human chromosome 19. Organization of genes for plasma lipid transport (APOC1, -C2 and -E and LDLR) and the genes C3, PEPD and GPI, *Proc. Natl. Acad. Sci. U.S.A.*, 83, 3929, 1986.
106. **MacKenzie, A. E., Korneluk, R. G., Zorzato, F., Fujii, J., Phillips, M., Iles, D., Wieringa, B., Le Blond, S., Bailly, J., Willard, H. F., Duff, C., Worton, R. G., and MacLennan, D. H.,** The human ryanodine receptor gene: its mapping to 19q13.1, placement in a chromosome 19 linkage group and exclusion as the gene causing myotonic dystrophy, *Am. J. Hum. Genet.*, 46, 1082, 1990.
107. **MacLennan, D. H., Duff, C., Zorzato, F., Fujii, J., Phillips, M., Korneluk, R. G., Frodis, W., Britt, B. A., and Worton, R. G.,** Ryanodine receptor gene is a candidate for predisposition to malignant hyperthermia, *Nature*, 343, 559, 1990.
108. **McCarthy, T. V., Healy, J. M. S., Heffron, J. J. A., Lehane, M., Deufel, T., Lehmann-Horn, F., Faralli, M., and Johnson, K.,** Localization of the malignant hyperthermia susceptibility locus to human chromosome 19q12-13.2, *Nature*, 343, 562, 1990.
109. **Otsu, K., Khanna, V. K., Archibald, A. L., and MacLennan, D. H.,** Co-segregation of porcine malignant hyperthermia and a probable causal mutation in the skeletal muscle ryanodine receptor gene in backcross families, *Genomics*, 11, 744, 1991.
110. **Gillard, E. F., Otsu, K., Fujii, J., Khanna, V. K., de Leon, S., Derdemezi, J., Britt, B. A., Duff, C. L., Worton, R. G., and MacLennan, D. H.,** A substitution of cysteine for arginine-614 in the ryanodine receptor is potentially causative of human malignant hyperthermia, *Genomics*, 11, 751, 1992.
111. **Gillard, E. F., Otsu, K., Fujii, J., Duff, C. L., de Leon, S., Khanna, V. K., Britt, B. A., Worton, R. G., and MacLennan, D. H.,** Polymorphisms and deduced amino acid substitutions in the coding sequence of the ryanodine receptor (RYR1) gene in individuals with malignant hyperthermia, *Genomics*, 13, 1247, 1992.
112. **MacLennan, D. H. and Phillips, M. S.,** Malignant hyperthermia, *Science*, 256, 789, 1992.
113. **Hogan, K., Couch, F., and Powers, P. A.,** A cysteine-for-arginine substitution (R614C) in the human skeletal muscle calcium release channel cosegregates with malignant hyperthermia, *Anesth. Analg.*, 75, 441, 1992.
114. **Otsu, K., Nishida, N., Kimura, Y., Kuzuya, T., Hori, M., Kamada, T., and Tada, M.,** The point mutation Arg^{615} to Cys in the Ca^{2+} release channel of skeletal muscle sarcoplasmic reticulum is responsible for hypersensitivity to caffeine and halothane in malignant hyperthermia, *J. Biol. Chem.*, 269, 9413, 1994.
115. **Quane, K. A., Healy, J. M. S., Keating, K. E., Manning, B. M., Couch, F. J., Palmucci, L. M., Doriguzzi, C., Fagerlund, T. H., Berg, K., Ording, H., Bendixen, D., Mortier, W., Linz, U., Müller, C. R., and McCarthy, T. V.,** Mutations in the ryanodine receptor gene in central core disease and malignant hyperthermia, *Nature Genet.*, 5, 51, 1993.
116. **Quane, K. A., Keating, K. E., Manning, B. M., Healy, J. M. S., Monsieurs, K., Heffron, J. J. A., Lehane, M., Heytens, L., Krivosic-Horber, R., Adnet, P., Ellis, F. R., Monnier, N., Lumardi, J., and McCarthy, T. V.,** Detection of a novel common mutation in the ryanodine receptor gene in malignant hyperthermia. Implications for diagnosis and heterogeneity studies, *Hum. Mol. Gen.*, 3, 471 1994.
117. **Quane, K. A., Keating, K. E., Healy, J. M. S., Manning, B. M., Krivosic-Horber, R., Krivosic, I., Monnier, N., Lunardi, J., and McCarthy, T. V.,** Mutation screening of the RYR1 gene in malignant hyperthermia. Detection of a novel Tyr to Ser mutation in a pedigree with associated central cores, *Genomics*, 23, 236, 1994.
118. **Rouquier, S., Giorgi, D., Trask, B., Bergman, A., Phillips, M. S., MacLennan, D. H., and de Jong, P.,** A cosmid and yeast artificial chromosome contig containing the complete ryanodine receptor *RYR*1 gene, *Genomics*, 17, 330, 1993.
119. **Roberts, R. G., Coffey, A. J., Bobrow, M., and Bentley, D. R.,** Exon structure of the human dystrophin gene, *Genomics,* 16, 536, 1993.

120. **Levitt, R. C., Nouri, N., Jedlicka, A. E., McKusick, V. A., Marks, A. R., Shutack, J. G., Fletcher, J. E., Rosenberg, H., and Meyers, D. A.,** Evidence for genetic heterogeneity in malignant hyperthermia susceptibility, *Genomics*, 11, 543, 1991b.
121. **Deufel, T., Golla, A., Iles, D., Meindl, A., Meitinger, T., Schindelhauer, D., DeVries, A., Pongratz, D., MacLennan, D. H., Johnson, K. J., and Lehmann-Horn, F.,** Evidence for genetic heterogeneity of malignant hyperthermia susceptibility, *Am. J. Hum. Genet.*, 50, 1151, 1992.
122. **Fagerlund, T., Islander, G., Ranklev, E., Harbitz, I., Hauge, J. G., Mekleby, E., and Berg, K.,** Genetic recombination between malignant hyperthermia and calcium release channel in skeletal muscle, *Clin. Genet.*, 41, 270, 1991.
123. **Saidman, K., Havard, E., and Egger, E.,** Hyperthermia during anesthesia, *J. Am. Med. Assoc.*, 190, 1029, 1964.
124. **O'Brien, P. J., Klip, A., Britt, B. A., and Kalow, B. I.,** Malignant hyperthermia susceptibility. Biochemical basis for pathogenesis and diagnosis, *Can. J. Vet. Res.*, 54, 83, 1990.
125. **Ball, S. P., and Johnson, K. J.,** The genetics of malignant hyperthermia, *J. Med. Genet.*, 30, 89, 1993.
126. **Levitt, R. C., Olckers, A., Meyers, S., Fletcher, J. E., Rosenberg, H., Isaacs, H., and Meyers, D. A.,** Evidence for the localization of a malignant hyperthermia susceptibility locus (MHS2) to human chromosome 17q, *Genomics*, 14, 562, 1992.
127. **Olckers, A., Meyers, D. A., Meyers, S., Taylor, E. W., Fletcher, J. E., Rosenberg, H., Isaacs, H., and Levitt, R. D.,** Adult muscle sodium channel α-subunit is a gene candidate for malignant hyperthermia susceptibility, *Genomics*, 14, 829, 1992.
128. **Gregg, R. G., Couch, F., and Hogan, K.,** Assignment of the human gene for the beta subunit of the voltage-dependent calcium channel (CACNLB1) to chromosome 17 using somatic cell hybrids and linkage mapping, *Genomics*, 15, 185, 1993.
129. **Iles, D. E., Segers, B., Sengers, R. C. A., Monsieurs, K., Heytens, L., Halsall, P. J., Hopkins, P. M., Ellis, F. R., Hall-Curran, J. L., Stuart, A. D., and Wieringa, B.,** Genetic mapping of the β1 and γ subunits of the human skeletal muscle L-type voltage dependent calcium channel on chromosome 17q and exclusion as a candidate gene for malignant hyperthermia susceptibility, *Hum. Mol. Genet.*, 2, 863, 1993.
130. **Fontaine, B., Khurana, T. B. S., Hoffman, E. P., Bruns, G. A. P., Haines, J. L., Trofatter, J. A., Hanson, M. P., Rich, J., McFarlane, H., Yasek, D. M., Romano, D., Gusella, J. F., and Brown, R. H.,** Hyperkalemic periodic paralysis and the adult muscle sodium channel α-subunit gene, *Science*, 250, 1000, 1990.
131. **Sudbrak, R., Golla, A., Powers, P., Gregg, R., Du Chesne, I., Lehmann-Horn, F., and Deufel, T.,** Exclusion of malignant hyperthermia susceptibility (MHS) from a putative MHS2 locus on chromosome 17q and of the α1, β1, γ subunits of the dihydropyridine receptor calcium channel as candidates for the molecular defect, *Hum. Mol. Genet.*, 2, 857, 1993.
132. **Gregg, R. G., Couch, F., Hogan, K., and Powers, P. A.,** Assignment of the human gene for the α 1 subunit of the skeletal muscle DHP-sensitive calcium channel (CACNL1A3) to chromosome 1q31-q32, *Genomics*, 15, 107, 1993.
133. **Lehmann-Horn, F., Jurkat-Rott, K., Elbaz, A., Heine, R., Melzer, W., Sipos, I., Gregg, R. G., Hogan, K., Powers, P. A., Lapie, P., Vale-Santos, J., Weissenbach, J., and Fontaine, B.,** A calcium channel mutation causing a human disease. Hypokalemic periodic paralysis, *Soc. Gen. Physiol. 48th Annu. Meet.*, 1994, 48, 5a,
134. **Iles, D. E., Lehmann-Horn, F., Scherer, S. W., Tsui, L. C., Weghuis, D. O., Suijkerbuijk, R. F., Heytens, L., Mikala, G., Schwartz, A., Ellis, F. R., Stewart, A. D., and Wieringa, B.,** Localization of the gene encoding the α2/δ-subunits of the L-type voltage-dependent calcium channel to chromosome 7q and analysis of the segregation of flanking markers in malignant hyperthermia susceptible families, *Hum. Mol.. Genet.*, 3, 969, 1994.

135. **Gyapay, G., Morrissette, J., Vignal, A., Dib, C., Fizames, C., Millasseau, P., Marc, S., Bernardi, G., Lathrop, M., and Weissenbach, J.,** The 1993-94 Généthon human linkage map, *Nature Genet.*, 7, 246, 1994.
136. **Shy, G. M. and Magee, K. R.,** A new congenital non-progressive myopathy, *Brain*, 79, 610, 1956.
137. **Patterson, V. H., Hill, T. R., Fletcher, P. J., and Heron, J. R.,** Central core disease: clinical and pathological evidence within a family, *Brain*, 102, 581, 1979.
138. **Ramsey, P. L. and Hensinger, R. N.,** Congenital dislocation of the hip associated with central core disease, *J. Bone Joint Surg.*, 57A, 648, 1975.
139. **Dubowitz, V. and Platts, M.,** Central core disease of muscle with focal wasting, *J. Neurol. Neurosurg. Psychiatry*, 28, 432, 1965.
140. **Shuaib, A., Paasuke, R. T., and Brownell, K. W.,** Central core disease: clinical features in 13 patients, *Medicine*, 66, 389, 1987.
141. **Dubowitz, V. and Pearse, A. G. E.,** Oxidative enzymes and phosphorylase in central-core disease of muscle, *Lancet*, 23, 1960.
142. **Isaacs, H., Heffron, J. J. A., and Badenhorst, M.,** Central core disease. A correlated genetic, histochemical, ultramicroscopic and biochemical study, *J. Neurol. Psych.*, 38, 1177, 1975.
143. **Byrne, E., Blumbergs, P. C., and Hallpike, J. F.,** Central core disease. Study of a family of five affected generations, *J. Neurol. Sci.*, 53, 77, 1982.
144. **Hayashi, K., Miller, R. G., and Brownell, A. K. W.,** Central core disease: ultrastructure of the sarcoplasmic reticulum and T-tubules, *Muscle Nerve*, 12, 95, 1989.
145. **Dubowitz, V. and Roy, S.,** Central core disease of muscle: clinical histochemical and electron microscopic studies of an affected mother and child, *Brain*, 93, 133, 1970.
146. **Eng, G. D., Epstein, B. S., Engel, W. K., McKay, D. W., and McKay, R.,** Malignant hyperthermia and central core disease in a child with congenital dislocating hips, *Arch. Neurol.*, 35, 189, 1978.
147. **Frank, J. P., Harati, Y., Butler, I. J., Nelson, T. E., and Scott, C. I.,** Central core disease and malignant hyperthermia syndrome, *Ann. Neurol.*, 7, 11, 1980.
148. **Haan, E. A., Freemantle, C. J., McCure, J. A., Friend, K. L., and Mulley, J. C.,** Assignment of the gene for central core disease to chromosome 19, *Hum. Genet.*, 86, 187, 1990.
149. **Kausch, K., Lehmann-Horn, F., Janka, M., Wieringa, B., Grimm, T., and Müller, C. R.,** Evidence for linkage of the central core disease locus to the proximal long arm of human chromosome 19, *Genomics*, 10, 765, 1991.
150. **Mulley, J. C., Kozman, H. M., Phillips, H. A., Gedeon, A. K., McCure, J. A., Iles, D. E., Gregg, R. G., Hogan, K., Couch, F. J., Weber, J. L., MacLennan, D. H., and Haan, E. A.,** Refined genetic localization for central core disease, *Am. J. Hum. Genet.*, 52, 398, 1993.
151. **Mignery, G. A. and Südhof, T. C.,** The ligand binding site and transduction mechanism in the inositol-1,4,5-triphosphate receptor, *EMBO J.*, 9, 3893, 1990.
152. **Miyawaki, A., Furuichi, T., Ryou, Y., Yoshikawa, S., Nakagawa, T., Saitoh, T., and Mikoshiba, K.,** Structure-function relationships of the mouse inositol 1,4,5-trisphosphate receptor, *Proc. Natl. Acad. Sci. U.S.A.*, 88, 4911, 1991.
153. **Chen, S. R. W., Airey, J. A., and MacLennan, D. H.,** Positioning of major tryptic fragments in the Ca^{2+} release channel (ryanodine receptor) of rabbit skeletal muscle sarcoplasmic reticulum, *J. Biol. Chem.*, 268, 22642, 1993.
154. **Carafoli, E.,** Intracellular calcium homeostasis, *Annu. Rev. Biochem.*, 56, 395, 1987.
155. **Wrogemann, K. and Pena, S. D. J.,** Mitochondrial calcium overload: a general mechanism of cell necrosis in muscle diseases, *Lancet*, 1, 672, 1976.
156. **Heiman-Patterson, T. D., Rosenberg, H. R., Binning, C. P. S., and Tahmoush, A. J.,** King-Denborough Syndrome: contracture testing and literature review, *Pediatr. Neurol.*, 2, 175, 1986.

157. **Dickson, D.,** DNA testing helps British bring better pig to market, *Nature*, 362, 688, 1993.
158. **MacLennan, D, H. and Chen, S. R. W.,** The role of the calcium release channel of skeletal muscle sarcoplasmic reticulum in malignant hyperthermia, *Ann. N.Y. Acad. Sci.*, 707, 294, 1993.
159. **Sudbrak, R. et al.,** Mapping of a further malignant hyperthermia susceptibility locus to chromosome 3q13.1, *Am. J. Hum. Gen.*, 56, 684, 1995.

INDEX

A

B

D

E

F

J

K

L

M

N

Q

R

T